WAVES OF KNOWING

Karin Amimoto Ingersoll

WAVES OF KNOWING

A Seascape Epistemology

Duke University Press ~ Durham and London ~ 2016

Designed by Courtney Leigh Baker
Typeset in Arno Pro by Copperline.

Library of Congress Cataloging-in-Publication Data
Names: Ingersoll, Karin E.
Title: Waves of knowing : a seascape epistemology /
Karin Amimoto Ingersoll.
Description: Durham : Duke University Press, 2016. |
Includes bibliographical references and index.
Identifiers: LCCN 2016007858 (print) | LCCN 2016008644 (ebook) ISBN
9780822362128 (hardcover : alk. paper)
ISBN 9780822362340 (pbk. : alk.paper)
ISBN 9780822373803 (e-book)
Subjects: LCSH: Hawaiians. | Surfing—Hawaii.
Classification: LCC DU624.65 I69 2016 (print) | LCC DU624.65 (ebook) |
DDC 305.899/42—dc 3
LC record available at HTTP://lccn.loc.gov/2016007858

Cover art: Abigail Lee Kahilikia Romanchak, *Watermark* (detail),
collagraph. Courtesy of the artist.

This book is dedicated to my two boys,
Kahoe King and Nāheʻolu Clive.
You both are the inspiration for my
heart's work and happiness.

Contents

Acknowledgments

I want to thank my family for their unwavering support. Mom and Dad, thank you for encouraging me to follow my dreams, and Kristin, for always believing. Russell, thank you for inspiring me and my work, and, more importantly, for always being there.

I wish to acknowledge the significant contributions to this work of my professors at the University of Hawaiʻi at Mānoa: Michael J. Shapiro, Jonathan Goldberg-Hiller, and Noenoe K. Silva. Your guidance and generous sharing of your always brilliant insight have made this work possible.

Many thanks too to Teweiariki Teaero for permission to reproduce his poem "Ocean Heart Beat" from Waʻa in Storms in the introduction, and to Robert Sullivan and Auckland University Press for permission to use "Waka 93" from Star Waka (1999) in the epilogue.

I also owe much gratitude to those who participated in oral histories, sharing thoughts and stories from their lives and from their important work: Bruce M. Blankenfeld, Nainoa Thompson, John R. Kukeakalani Clark, Duane DeSoto, Kanekoa Schultz, Angela Hiʻilani Kawelo, and Vicki Holt-Takamine.

Mahalo also to Cody Pueo Pata and Keoni James Kuoha for editing the Hawaiian language and content in this book. Me ka haʻahaʻa.

Indigenous Terrain

When I enter the ocean, my indigenous identity emerges. I become a historical being riding waves, running as a liquid mass, pulled up from the deep and thrown forward with a deafening roar. I disappear with fish and strands of seaweed as I course through veins of ocean currents. John Muir spoke of how he went out for a walk and stayed until sundown. "For going out," he said, "I found, I'm really going in." When I enter the sea, I enter a process of reimagination as the power of the ocean continually reshapes me alongside the coastal shores of my home.

Hitting that first whitewall of water, I become a Kanaka Maoli (Native Hawaiian) surfer. I ride waves; read the wind, swell directions, and tides; know the reefs and the seasonal sand migrations; and find myself most comfortable floating atop a board with my *na'au* (gut), mind, and heart facing the sea. In *ma ke kai* (in the sea), my physical involvement with, and thus my physical capabilities in, the world evolve. I become more agile in the water than on land: I can soar, glide, dive, and spin. I'm faster in the ocean, and can better navigate coral heads than roads. Sounds, smells, and tastes expand to include those not found on terra firma. I become aware of my pelagic origin as I soak in the same salty waters as Kanaka Maoli centuries before me.

It isn't until I enter ke kai for *he'e nalu* (surfing) that I am able to reconnect with my Kanaka heritage. On his deathbed in 1972, my mother's father revealed his Kanaka blood to her with tears in his eyes. His family tried to cover up that his grandmother was a Hawaiian woman because at that

FIGURE I.1. Karin Amimoto Ingersoll, "Becoming," 2009.
Photograph by Russell J. Amimoto.

time in the Chinese community, it was shameful to be anything other than
Chinese—particularly Kanaka. I have no legal documentation of my Hawaiian blood, leaving me with only the oral history and photographs passed
on to me by my mother. Questions of legitimacy ring loudly in Hawaiian
circles: How much Hawaiian are you? Where is your ʻohana (family) from?
What is your Hawaiian name? Do you have documentation? I accept my
mother's oral history as authentic and legitimate, and am aware that identity is inherently political and dynamic. Many people in Hawaiʻi encourage
me to embrace my oral history, often sharing similar colonial stories of lost
identity and documentation, but I am aware that my "Hawaiian" identity is
illegitimate to many other individuals and institutions. I remain forever in-
between. I am an extreme case of cultural and genealogical dilution, and yet
I believe that my circumstance has offered me a specific place from which to

analyze the political and philosophical power of ke kai, a power that I feel each time I enter the sea.

As I build upon the burgeoning effort in academia toward the retrieval of local ways of knowing, researching, and producing knowledge, the West is not the axis of negation that moves my articulations and reactions, because in a multisited world, our intelligibility is an interconnected matrix. Instead, my aim is to pull indigenous peoples away from the binary oppositions between the "colonizer" and the "colonized," to minimize the "otherness" from both sides, and to decenter the conversation toward independent and alternative ways of knowing and producing knowledge that allow for empowerment and self-determination within a modern and multisited world.[1]

I write this book as a Kanaka Maoli surfer sitting within a colonial landscape, discovering how ke kai enables an autonomous reconnection, recreation, and reimagination for all Kanaka Maoli through an ocean-based epistemology. This work articulates the potential power that he'e nalu and other ocean-based knowledges, such as ho'okele (navigation) and lawai'a (fishing), offer to cultural awareness and affirmation within the reality of a colonial history and of neocolonial systems that continue to subject Hawaiian knowledges and identities. Ocean-based knowledges help to mobilize Hawaiian bodies through a specific time and space in ke kai that anchors Kānaka [plural of "Kanaka"] in existence as ever-shifting and negotiating beings within the Western institutions of statehood, capitalism, and ecologically challenging development. This work aims to bring the physical movements of he'e nalu, ho'okele, and lawai'a, back into an ontological perspective that speaks to an ethical experience of movement through the world and life.

I first realized the sea's profound potential when I was hired to act as a surf guide for the surf camp Sa'Moana on the island of Upolu in Samoa. When I left for Samoa, O'ahu's waters were increasingly congested with tourists and new surfers lured into the fantasy of a surf lifestyle. Thus, the opportunity to expand my surf territory to a South Pacific destination where I could continue to challenge my fears, my knowledge of geography, geology, and astronomy, and to enhance my connection with the ocean's moods, had strong siren-like effects on me, as it would any surfer. The first few days on Upolu fulfilled my every prefabricated fantasy. I would wake before dawn, roll out from under my mosquito net, emerge from my fale (Samoan thatched house), and peek out to check the palm fronds. No movement—a good sign for top surfing conditions. We would load the truck with surf gear, prepacked lunches, and our surf guests, and head off to the ocean for a pristine session. I was up, moving, and gliding across water as the sun only began to cast its first beams of pink and orange over the coconut trees. This was the most alive I had ever felt.

But the dynamite used to bomb the brilliant green and blue Samoan reefs (sustenance for local Samoans and for innumerable marine species) to pave a channel for the camp's surf boats shook my callow mind-set. The brutal reef destruction caused profound economic, cultural, environmental, and spiritual disruption because Samoan culture and sustainability is inextricably linked to the ocean, its coral, and fish. The connection is sacred. There were other acts of colonization taking place within my surf dream too: the low wages of the Samoan staff, the culturally condescending relationships between the white Australian owners and the local staff, the camp operators' refusal to speak or learn Samoan, and ruptures in Samoan political and social systems caused by the camp's presence and conduct. For instance, the surf camp created a physical and cultural barrier between the camp on the beach and the surrounding village by erecting gates and manipulating vegetation to create a sense of "private property" and segregation. The land that the surf camp sits on is rented, not owned. The Samoan government has been prudent enough not to allow land sales to anyone not born in Samoa, yet foreign capitalists have continued to find cracks to shove their fingers into, tearing open as large a gap as possible for monetary outflow.

I began to see underlying political, social, and ethical issues hiding within my fantasy that had been glossed over by surfing films, magazines, advertisements, and the tourism industry. We, as surfers, had been told by the mass media that to travel to surf destinations such as Sa'Moana was to build upon one's "authentic" surf identity as a "soul" surfer who lives to explore and experience the world's oceans. The potential cost of this sojourning, however, is omitted from the narrative, as is an awareness of the role that surf camps, magazines, advertisements, clothing, and equipment companies play in the capitalization of these ocean locations. I began to see the impact our desire to ride waves had on those island locales where we desired to experience them. Despite our perceived identities as organic beings, surfers are neither innocent nor benign voyagers, and our experiences and our practices often escape our intentions and philosophies. Surfers are no longer merely a community of stereotypical antiestablishment thrill seekers, we are now also a group of international, neocolonial capitalists "discovering" new waves in Oceania (and elsewhere).

Reflecting on the effects of the surf tourism industry in Samoa, I wondered how surf tourism was affecting Kānaka Maoli in Hawaiʻi. Surf tourism is much more developed in Hawaiʻi than it is in Samoa due to the islands' geographic proximity to the U.S. mainland and the political circumstance of Hawaiʻi being a colonized nation. Furthermore, surfing survived the process of cultural colonization in Hawaiʻi, which it did not do in Samoa, making the activity much more predominant in Hawaiʻi today. Most significantly, however, it is the geographic makeup of the Hawaiian Islands, which naturally and consistently produces ideal waves, that has rewarded Hawaiʻi with some of the best surfing in the world. Hawaiʻi has become a modern surf utopia.

Under this international label of a surf paradise I wondered about the impact that surf tourism (which began in the 1920s) has had and continues to have on the islands and indigenous people of Hawaiʻi. How is the surf industry part of a neocolonial project of domination in Hawaiʻi? Kānaka Maoli have been struggling to decolonize after a two-hundred-year-long epistemological, ontological, political, economic, and geographic colonization by the United States; what role does this evolving industry play in the larger project of Western domination? More significantly, how does Hawaiian heʻe nalu, along with other oceanic activities, simultaneously serve as a means of empowerment for Kānaka Maoli within the proliferating surf tourism industry?

I have found that the burgeoning neocolonial surf industry also offers Kānaka empowerment through the indigenous enactment of heʻe nalu because the oceanic literacy of heʻe nalu, as well as of hoʻokele, lawaiʻa, luʻu (diving), *hoe waʻa* (canoe paddling), and other activities, all involve a knowledge situated in a specific place and space, which is oceanic. These enactments involve a knowledge that reconnects Kānaka to our pasts and to our ancestors as understood through our oral histories. They involve a literacy that empowers Native Hawaiians because they offer self-sufficiency, honor Kanaka native histories, allow for flexibility and movement, and offer philosophical nourishment for visions of alternative and self-determined futures. I found that these literacies involve a Kanaka epistemology, an oceanic knowledge that privileges an alternative political and ethical relationship with the surrounding physical and spiritual world.

I have termed this epistemology "seascape epistemology." It is an approach to knowing presumed on a knowledge of the sea, which tells one how to move through it, how to approach life and knowing through the movements of the world. It is an approach to knowing through a visual, spiritual, intellectual, and embodied literacy of the ʻāina (land) and kai (sea): birds, the colors of the clouds, the flows of the currents, fish and seaweed, the timing of ocean swells, depths, tides, and celestial bodies all circulating and flowing with rhythms and pulsations, which is used both theoretically and applicably by Kānaka Maoli today for mobility, flexibility, and dignity within a Western-dominant reality. Seascape epistemology embraces an oceanic literacy that can articulate the potential for travel and discovery, for a re-creation and de-creation. Seascape epistemology also allows us to produce our own bodies of scholarship in a colonial reality that has rendered Native Hawaiian knowledge "cultural" rather than intellectual or academic. It helps to create a paradigm for relocating Hawaiian identity back into ke kai.

As a philosophy of knowledge, seascape epistemology does not encompass a knowledge of "the ocean" and "the wind" as things. Seascape epistemology is not a knowledge of the sea. Instead, it is a knowledge about the ocean and the wind as an interconnected system that allows for successful navigation through them. It's an approach to life and knowing through passageways. Seascape epistemology organizes events and thoughts according to how they move and interact, while emphasizing the importance of knowing one's roots, one's center, and where one is located inside this constant movement. As Kānaka travel, modernize, and adapt as multisited and complex individuals, seascape epistemology enables us to observe and interpret diverse knowledges from our own native perspectives. The power of seascape epistemology lies in its organic nature, its inability to be mapped absolutely, and its required interaction with the intangible sea.

In articulating how Hawaiian knowledge intersects dominant narratives and systems, I look into ke kai because Kānaka Maoli have a uniquely *Hawaiian* relationship to the ocean that is moored in a historical relationship—in which the ocean serves as an instrument of migration; as transportation; and as a source of food, medicine, and shelter—as well as spiritual right and responsibility, or *kuleana*, to the sea expressed in the concept of *mālama ʻāina* (caring for the land). The ocean is where we cleanse, dance, play, train, and die. It is the point from which we have always leapt off, physically and philosophically, into our pasts and our futures. It is also the pathway that brought our colonizers to us: captain James Cook, missionaries, foreign merchants, whale fisheries, and the subsequent naval ships from Britain, France, and the United States, all vying for power and influence in Oceania. It is the pathway that brings destruction through tsunamis, hur-

FIGURE I.2. "The Intangible Seascape," 2009. Photograph by Russell J. Amimoto.

ricanes, and drowning. Yet it is that same pathway that connects us in a familiar constellation of islands to our Oceanic neighbors, the world, and the cosmos beyond the horizon.

A Hawaiian relationship to *ka moana* (the open ocean) is also genealogical, and Kānaka Maoli attuned to this historical ontology believe that our essence, as a people, is born from the sea. Not only our "identity" as political and social beings but our very being-in-the-world and being-in-time comes out of the ocean. It is *He Kumulipo*, a predominant genealogy *oli* (chant) composed by a Kanaka priest around the eighteenth century, that determines this ontological connection to the sky, ocean, cosmos, plants, animals, and land.[2] The oli narrates that darkness, or Pō, spontaneously gave birth to a son, Kumulipo, and a daughter, Pō'ele, and that these two in turn gave birth to the coral polyp in the sea, and many other creatures followed, first in the sea and then on land. Kanaka scholar Lilikalā Kame'eleihiwa, explains, "In the *Kumulipo*, Hawaiian time begins with darkest night, the ancient female ancestor, who gives birth to male and female nights. Brother and sister mate in an incestuous union to produce the divinity of the universe, which is all life. They give birth to the coral polyp in the fundamental slime of the earth; each creature in its turn gives birth to other sea creatures and seaweeds, proceeding up the evolutionary chain through the fish, birds, and creeping things . . ." (Kameʻeleihiwa 1992, x)

Hawaiian origin, then, is within the ocean. Kameʻeleihiwa continues, "Hawaiian mythology recognizes a prehuman period before mankind was born when spirits alone peopled first the sea and then the land, which was born of the gods and thrust up out of the sea" (5).

The genealogy continues with the creation of the Hawaiian people, connecting Hawaiian origin to the space and place of the sea.

Kameʻeleihiwa asserts, "Hawaiian identity is, in fact, derived from the *Kumulipo*, the great cosmogonic genealogy. Its essential lesson is that every aspect of the Hawaiian conception of the world is related by birth, and as such, all parts of the Hawaiian world are one indivisible lineage" (2). This fact affects every aspect of knowledge and sense of being for Kānaka Maoli. Kanaka ancestors believed that human beings were a part of not only the sea but the universe; the ocean was the essence of their own identity.

Geologically, all Pacific Islands were born up out of the sea, linking the land to the ocean physically, genealogically, and metaphorically. This notion of "Mother Sea" is not isolated to Oceania. Western science has strong evidence to support the notion that human beings evolved from microorganisms in the sea. Environmentalist Rachel Carson writes, "I tell here the story of how the young planet Earth acquired an ocean. . . . The story is founded on the testimony of the earth's most ancient rocks, . . . on other evidence written on the face of the earth's satellite, the moon; and on hints contained in the history of the sun, and the whole universe of star-filled space" (Carson 1989, 3).

Carson continues to explain how the moon is arguably a child of the Earth, which was then a great mass of molten liquid experiencing tremendous tides dictated by the pull of the sun. The moon is said to be a "great tidal wave" torn off the Earth and hurled into space. The scar or depression left by this great wave holds the Pacific Ocean. Carson states,

As soon as the earth's crust cooled enough, the rains began to fall. . . . And over the eons of time, the sea has grown even more bitter with the salt of the continents. . . . It seems probable that, within the warm saltiness of the primeval sea, certain organic substances were fashioned from carbon dioxide, sulphur, nitrogen, phosphorus, potassium, and calcium. Perhaps these were transition steps from which the complex molecules of protoplasm arose—molecules that somehow acquired the ability to reproduce themselves and begin the endless stream of life. (7)

Alongside and predating the facts and findings of Western science concerning human evolution and the ocean, Oceanic histories, religious objects, and rituals have always narrated the genealogical (evolutionary) relationship between the sea and human beings. From these historical claims emerge a Kanaka Maoli ontology and epistemology that is intertwined with the sea, that is bound by diversity yet remains a functioning whole. Kānaka Maoli exist and "know" through an interaction with the sea; it is a genealogical engagement with Hawaiian ancestral parents, Papahānaumoku (Earth Mother) and Wākea (Sky Father), who are an embodiment of the ʻāina. Kameʻeleihiwa explains that the two were half siblings by the ʻŌpūkahonua lineage and mated to give birth to the islands as well as Hoʻohōkūkalani, their first human offspring (Kameʻeleihiwa 1992, 25). Kanaka knowledge is a totality of everything as an intertwined lineage, and knowing speaks to a personal knowledge embedded in a specific history, culture, and time that is reactivated, in part, for contemporary Kānaka Maoli like me through oceanic enactments such as heʻe nalu, hoʻokele, and lawaiʻa.

Sitting on my surfboard, bones moving with currents, I think about how a reimagination of my indigenous identity, one re-created in a modern world, for a modern and multisited Kanaka, is anchored in historical culture. I think of how Hawaiian royalty, aliʻi (chiefs), distinguished themselves and their superior ability to ride waves with personal ocean songs and chants. These oli acted as proclamations of grandeur and expertise, combining historical perceptions of place with an individual's personal response, experience, and connection to that place. Kanaka scholar and cultural expert Mary Kawena Pukui narrates a section of a surf chant written about Naihe, an aliʻi from the district of Kaʻū, on the island of Hawaiʻi (the Hawaiian text was not available):

The great waves, the great waves rise in Kona,
Bring forth the loin cloth that it may be on display,
The ebbing tide swells to set the loin cloth flying,
The loin cloth, Hoaka, that is worn on the beach,

It is the loin cloth to wear at sea, a chief's loin cloth,
Stand up and gird on the loin cloth
The day is a rough one, befitting Naihe's surfboard,
He leaps in, he swims, he strides out to the waves,
The waves that rush hither from Kahiki.
White capped waves, billowy waves,
Waves that break into a heap, waves that break and spread.
The surf rises above them all,
The rough surf of the island,
The great surf that pounds and thrashes
The foamy surf of Hikiau,
It is the sea on which to surf at noon,
The sea that washes the pebbles and corals ashore.
(Pukui 1949, 256)

A wave for heʻe nalu becomes a cultural resource. The "great waves" roll in to Kona on the island of Hawaiʻi from Kahiki, (the East, and, by extension, all foreign lands; also, but not exclusively, Tahiti), born from a great cultural and genealogical origin. These waves also bring forth the loincloth of the aliʻi; supporting and propelling the aliʻi, the marrow of Kanaka society, yet the surf "rises above them all." The waves "peak and spread," washing pebbles and corals ashore, and delivering a divine and powerful knowledge and way of being to the people of Hawaiʻi. Naihe's oli reinforces the image and idea of the physical structure of the wave as a way of knowing the past, and thus a way of understanding of the present within that context. Naihe's oli illuminates how the act of entering ke kai today, to surf, voyage, fish, dive, or swim, brings about a space and time that anchors Kānaka Maoli in a historical existence as ever-shifting and negotiating beings. Seascape epistemology privileges this Kanaka ontological connection to ke kai.

My work develops this oceanic potential by articulating the local ways in which contemporary Kānaka Maoli construct and conceive knowledge through ke kai. I explore how Kanaka empirical experiences encountered in the world, experiences such as heʻe nalu, connect to a politicalization of Hawaiian knowledge and place that involves interconnection, flexibility, and movement. There is a potentiality for Kānaka Maoli to turn ocean-based knowledges and practices into a politics and ethics (oceanic literacy is not political or ethical on its own) because ka moana is mixed into Native Hawaiian genealogy, place, and history, and thus helps lay the foundation for articulating Kanaka theories and other ways of knowing that arise from a connection to ʻāina. As a political and ethical act, heʻe nalu re-creates patterns already set in motion by our ancestors.

The colonial history of the Hawaiian Islands and its people has been thoroughly documented by scholars analyzing, for instance, the great land divide of 1848 called the Great Māhele, the 1897 petition against American annexation, the 1898 annexation, and gaining American statehood in 1959. These scholars and artists have brilliantly unveiled the many and deep consequences stemming from these, and other, pivotal events. Many layers of this living history have, and continue to be, peeled back, illuminating how American colonization directly affects Hawaiian health, wealth, dignity, culture, language, spirituality, and land today. Loss of land is a particularly critical element of Hawaiian colonization (and decolonization) because Kanaka language, economy, politics, and culture are all connected to ʻāina, both land and sea.

Recognizing that land-based literacy is as critical for Kānaka Maoli today as it was historically, I look to the sea in this analysis because, in addition to the sea's symbolic power as a life-sustaining body that connects the world's continents, I believe that it has become an increasingly contested (and environmentally distressed) space in Hawaiʻi. The profound effects of the colonization of land in the Hawaiian Islands necessitates a critical analysis of the current attempts to colonize the sea, which can be seen through the privatization of oceanfront areas, the lack of public access to certain coastal regions, developmental and agricultural runoff (pesticides and fertilizers), mass commercial fishing, the establishment of state zones in the sea, and, most significantly for this work, the emergence of a burgeoning surf tourism industry that has grown out of the larger Western colonial project in Hawaiʻi by presenting ke kai as a place of consumption and conquest for the surf tourist.

Neocolonial Surf Tourism in Hawaiʻi

"Surf tourism," according to Dr. Martin Fluker of Victoria University, may be defined as the act of traveling to either a domestic location for a period of time not exceeding six months or an international location for a period of time not exceeding twelve months and staying at least one night, with the primary motivation for selecting the destination being to actively participate in the sport of surfing (where the surfer relies on the power of the wave for forward momentum) (Fluker 2002). Following this definition, a self-defined surfer from Iowa who stays in Hawai'i for longer than six months is suddenly no longer a tourist. Does this person transform into a "local"? Kanaka spaces within the surf tourism fantasy get muddled when blended into this single, dominant, and stringent definition of the industry, which restricts the expansive potential of these spaces to be both Kanaka *and* tourist. For example, Kānaka Maoli in Hawai'i travel across and into tourist spaces by participating in local surf contests, advertisements, and hotel employment. Fluker's definition swallows up Kanaka spaces in statistical maps.

Ralph Buckley, director of the International Centre for Ecotourism Research, School of Environmental and Applied Sciences, Griffith University, in Queensland, Australia, adds to Fluker's definition: "In terms of economic statistics, surfing becomes tourism as soon as surfers travel at least 40km and stay overnight with surfing as the primary purpose for travel" (Buckley 2002, 407). These two definitions of surf tourism are Western-oriented and fail to consider how Kānaka Maoli travel on the seascape. In Buckley's definition, a geographic scale of distance is honored as the determining factor for a local versus a tourist identity. Following this, a Kanaka Maoli who lives in Hilo and drives to Kona to surf and stay overnight at the house of a relative or *hānai* (adoptive/calabash) family would be designated a tourist.

A Kanaka-centered definition of surf tourism might be the act of a person, event, or advertisement traveling to any destination outside that person's, event's, or advertisement's defined homeland with the main purpose of surfing for any period of time. Under this definition, Kānaka Maoli still find Kanaka spaces within the tourist industry, because in Hawai'i, Kānaka Maoli can be indigenous within a tourist space while not being tourists. The Hawaiian body and its relationship to the 'āina becomes a critical relationship of power for the Kanaka surfer within the neocolonial reality that surf tourism has created.

The evolution of surf tourism is directly tied to the larger project of political and economic colonization in Hawai'i. After the 1898 annexation of Hawai'i to the United States, American businessmen needed to present the islands to their fellow citizens as a valuable and desirable place, one of "soft primitivism." The activity of he'e nalu became a new commodity, a romanticized, chic, and adventurous selling point for Hawai'i. Today, this narrative has exploded into the international surf tourism arena, designating Hawai'i as *the* pleasure zone for this burgeoning, multibillion-dollar market. Mass media has established a global identity for surfers around which an economic and political agenda of governance has flourished. A new tactic of power has been systemically imposed on individual surfers, meticulously measuring the differences between their lives and that of an ideal surfer by imposing homogenous standards and expectations. The industry has rendered as enviable a surfer's designated motivations, spirituality, passions, and physical skills (Alcoff 2005). New modes of expertise have emerged and expanded as knowledges developed in the surfing industry through mass production: professional surfer, surf sponsor, surf journalist, photographer, surf coach, surf tour guides. A formulated surf industry has been established that embraces an ideology of consumption in which the surf tourist body is situated outside the places it visits, distancing the tourist from responsibility to, or respect for, the history and politics of these places. Native Hawaiian culture is not acknowledged as autonomous but is swallowed by a narrative of escapism and discovery: ocean places have been renamed in Hawai'i, and oceanic oral history has been overshadowed.

This book analyzes how surf tourism represents the problem that arises when state interests of power converge with capital interests of power, resulting in the violent marginalization and erasure of a people. I explore how the surf tourism industry perpetuates the dominance of a totalizing ideology that places indigenous identities, knowledges, imaginations, and memories in the periphery. Geographic and economic colonization is perpetuated, as is the specific Western epistemology regarding Hawai'i's role in a capitalist endeavor. The sea itself becomes the focal point of colonization by the industry that ideologically established ke kai as a place of conquest and domination.

Predominant surf tourism ethics and narratives that weave together histories of colonialism, militarism, and tourism, however, are not absolute. Kānaka Maoli not only move within the industry in beneficial ways but also help to shape the industry. For instance, Duke Paoa Kahinu Mokoe Hulikohola Kahanamoku, the most famous of the Waikīkī Beachboys, played a critical role in the development of surf tourism during the 1920s and '30s. His significant role in selling the pastime and islands as a commodity to the tourism industry established him not only as the "father of modern surfing" but also as the "ambassador of aloha." Duke officially introduced he'e nalu to Americans on the mainland United States while in Southern California for swimming exhibitions and meets in 1912 and again in 1916. His impact on the development of Australian surfing after his visit to the country in 1915 was also profound (Finney and Houston 1996). Yet as a beachboy, Duke did not just cater to tourist desires; he remained connected to ke kai during the turbulent time when global capitalism flooded the islands. His Kanaka values and knowledge of and respect for the ocean never faltered, even as he played the international role of an iconic beachboy.

The Waikīkī Beachboys were lifeguards, instructors, bodyguards, entertainers, and tour guides providing surf instruction and canoe rides for tourists fulfilling the surf-tourism narrative. Through the 1920s and '30s, the beachboys grew in popularity and their business ventures expanded. Kanaka scholar Isaiah Helekūnihi Walker asserts that the beachboys were not playing into the "tourist expectations of sexual conquest" but acting instead as anticolonial voices working to preserve "their surfing, culture, space and Hawaiian identities" (Walker 2008, 105). This is a critical point. Walker stresses the notion that Kanaka men (and women) developed their indigenous identities in the surf, and often "thwarted colonial encroachment" by transgressing colonial expectations and categories of what it meant to be "Hawaiian" (89). Walker argues that the beachboys, while catering to tourist business, were simultaneously defying colonial hegemony by retaining control over their aquatic domain. The *po'ina nalu* (the surf zone), not the beach or deep sea, but the wave zone itself, acted as place of autonomy, resistance, and survival for Kānaka. The po'ina nalu, as Walker argues, was (and is) a Hawaiian *pu'uhonua* (place of refuge) from colonial imposition and dominance. Walker explains, "In such a pu'uhonua, identities could be constructed in opposition to colonialism. This is not to say that colonialism had no influence on the shaping of such identities. Rather, Native Hawaiian identities fostered in the surf zone were developed in contrast to the colonial conquest on the shore. And, as a large part of this terrestrial conquest involved emasculating Native men . . . , the po'ina nalu was a location where Hawaiian men redefined themselves as active agents, embodying resistant masculinities" (92).

The Waikīkī Beachboys, for instance, officially organized themselves in 1911 into the Hui Nalu surf club to preserve he'e nalu from an exploitative *haole* (foreign, introduced) constituency that in 1908 had formed a whites-only surf club, the Outrigger Canoe Club, that boasted supremacy over Kānaka in Waikīkī waves. Kanaka surfers were not, however, submissive or "ideal natives," but were instead dominant and respected watermen who successfully established autonomous identities in opposition to colonial institutions and categories (105).

Kānaka Maoli have resisted commodification, and continue to do so, both inside and outside of the tourism industry, extracting historically and culturally based empowerment from heʻe nalu, as well as other ocean-based knowledges. Kānaka Maoli have inserted their own agency into what I deem a "neocolonial industry"; they are not only effected by the industry, they also affect it, making surf tourism a complex interaction rather than an imposition (Teaiwa 2001). Kānaka have always made new indigenous spaces within dominant structures, creating a politics that reaffirms Hawaiian identity while demanding participation and recognition in the system.

Surf tourism is a complex interaction between actors, ideologies, and events. Within this proliferating neocolonial industry, the indigenous activity of heʻe nalu provides itself as a means of empowerment specifically for Kānaka Maoli. I use the surf tourism industry to contextualize the significance of this ocean-based knowledge, as well as those of hoʻokele and lawaiʻa, as political and empowering by creating new spaces within dominant systems while demanding participation and recognition in the system. For Kānaka Maoli, the ocean becomes a place to re-create, regenerate, reaffirm, and return to autonomous and complex identities that are both historic and modern, both rooted and traveling. This is the potential of seascape epistemology: to help relocate modern Kanaka identities and bodies back into ke kai.

Why the Seascape?

A study of an indigenous epistemology reveals a homology between the treatment of epistemology and an analysis of the ocean as something fragile and changing, to an analysis of the ocean. There has been a predisposition of cultural and indigenous studies to connect indigeneity with territory, a "territory" that has been predominantly, although not entirely, land-based. My contribution speaks to the fluvial addition to the territorial through the Hawaiian seascape as a means of obtaining a geopolitical mapping of the political. Seascape epistemology dives into the ocean, splashing alternatives onto the Western-dominant and linear mind-set that has led the world toward realities of mass industrialization and cultural and individual assimilation. Understanding knowledge as an always moving interaction through theoretical frames challenges dominant theoretical narratives that strive to determine absolute "truths." This is the aim of seascape epistemology.

The seascape evokes powerful imagery as a place of adaptation, representing change, process, the inward and outward flows of ideas, reflections, and events. Water is the only chemical compound found as a solid, a liquid, and a gas, and it is both an acid and a base (Farber 1994). Water is multistructural as a formless phenomenon, yet it never loses its identity. In the same way, Hawaiian ways of being-in-the-sea transcend the physical world to include the metaphysical, spiritual, and sensational, creating codes of grammar through seascape epistemology, which normalize an indigenous sense of knowing and being that travels. Seascape epistemology enables a reading and a knowledge of the self that resists the petrification of its own dynamic character. Identity is always plural and in continual re-creation within seascape epistemology. It helps to mobilize Hawaiian bodies through a specific time and space in ke kai that anchors Kānaka in existence as ever-shifting and negotiating beings within a Western reality of statehood, capitalism, and ecologically challenging development that has altered Hawaiian epistemology.

Seascape epistemology builds off of Tongan scholar Epeli Hau'ofa's re-imagination of the ocean as a highway that links rather than separates Pacific Islanders in a "sea of islands." Hau'ofa offers a conception of culture, identity, and space that moves beyond what Banaban and I-Kiribati scholar Teresia K. Teaiwa describes as "discrete boundaries and disconnectedness" (Teaiwa 2005, 23). Hau'ofa's sea of islands defies the barriers established by development "experts," aid agencies, colonial governments, scientists, and select scholars, urging Pacific Islanders to instead decolonize our minds and recast our senses of identity by rediscovering the vision of our ancestors for whom the Pacific was a boundless sea of possibilities and opportunities (D'Arcy 2006, 7). Teaiwa asserts, "By emphasizing traditions of migration and voyaging, a matter-of-fact—if not fearless—approach to confronting difference, and the maintenance of kinship across vast distances, Hau'ofa's Oceanic peoples are exemplary of a Native way of being that is fluid, multiple and complex" (Teaiwa 2005, 23). Hau'ofa's concept allows not only Pacific identity to be fluid, multiple, and complex, but also Pacific epistemology.

This Oceanic concept helps develop a Kanaka relationship to a specific ontology and epistemology related to ke kai that resists colonial narratives, and that in turn helps to further develop my concept of seascape epistemology. I draw this connection based on the shared history of migration and cultural exchange in the region, a connection to which Hau'ofa speaks. This relationship is also expressed by Kānaka Maoli who refer to Pacific Islanders as "brothers" and "sisters" and reference ways of life in other Oceanic nations as applicable or similar to those in Hawai'i. I do recognize, however, that each Oceanic nation has a unique history, culture, language, and geography that should not be universalized. My argument is that the region shares many values and ambitions related to the common sea that surrounds and has profoundly shaped our cultures. It is this relationship that I lean upon as I exchange Pacific and Kanaka theories and ethics.

Aware of a critique on the problem of essentialization through such a regional vision, Hau'ofa recognizes that "our diverse loyalties are much too strong to be erased by a regional identity and our diversity is necessary for the struggle against the homogenizing forces of the global juggernaut. It is even more necessary for those of us who must focus on strengthening their ancestral cultures against seemingly overwhelming forces, to regain their lost sovereignty" (Hau'ofa 2005, 33–34). Not all of Oceania embraces the seascape in the same way, with the same practices, or with the same intensity. Margaret Jolly points out that Hau'ofa tends to essentialize a connection to and relationship with the ocean by all Pacific Islanders. She states, "For him the sea is as much inside the bodies of Islanders as it is their connecting fluid of passage, in world-traveling canoes or jumbo jets, the still center of an ocean of experience they navigate" (Jolly 2001, 419). Jolly challenges this perspective by stating that many Oceanic peoples today are not connected to the ocean geographically, economically, or psychologically. She also challenges the notion that Oceanic peoples are simply rooted as grounded in the land and becoming static in place and time, in boundaries of tradition, while foreigners travel, discover, and develop. It's a dialectical tension between movement and settlement, between routes and roots.

James Clifford, professor at the University of California, Santa Cruz, also warns of re-creating binary oppositions between indigenous notions of home and away, of the complex dynamic of "local landedness and expansive social spaces," in an attempt to articulate the full range of ways to be "modern." He stresses the importance in recognizing, "patterns of visiting and return, of desire and nostalgia, of lived connections across distances and differences" (Clifford 2001, 470).

Jolly criticizes Hau'ofa's celebration of the notion of a contemporary "world traveler," whom she argues is "more cramped—they typically follow older colonial circuits and their journeys are plotted by the cosmology of global capitalism" (Jolly 2001, 422). Furthermore, she argues that economics, geography, nation-state borders, and diaspora have hindered many islanders' ability to travel at all. She points out that Papua New Guinea, the

Solomon Islands, and Vanuatu don't feel a sense of ancestral connection to the ocean.

Essentializing an ocean-based relationship in Oceania, and more specifically in Hawai'i, is irresponsible and unproductive given the diverse cultures, economies, diasporic histories, and geographic locations of many Kānaka Maoli. I also recognize that not all Kānaka embrace ke kai in the same way, with the same practices, or with the same intensity. While this book situates Hawaiian bodies within a locatable spatial practice, it does not assume that an oceanic relationship is one accessible or familiar to all contemporary Kānaka 'Ōiwi, (Indigenous Peoples) nor that it is not. It does argue, however, that oceanic literacy is an ancestral knowledge historically relevant in Kānaka Maoli futures. All Kanaka ancestors sailed to the Hawaiian Islands, and thus in the goal of psychological decolonization the concept of knowledge in relation to the ocean and the surf is at least metaphorically applicable for all Kanaka Maoli. Hawaiian culture always meanders back to the sea. Ke kai is referenced in relation to Kanaka art, architecture, love, political power, spiritual awareness, and ancestry. Seascape epistemology builds upon this relationship, offering contemporary Kānaka Maoli alternative ways of navigating through the world, regardless of our current "places" in it. The concept of the seascape is valuable beyond the literal place of the sea. It is also a Kanaka discourse that tells us how to move through life as Kānaka have always moved, physically, culturally, and intellectually (Teaiwa 2005). The concept of a seascape honors this "travel," growth, and change through an approach to knowing that embraces that which is historically oceanic.

The place of ke kai is both constant and fluid, always changing and moving, like identity, and like our bodies. There is an essentialist relationship in seascape epistemology because it incorporates embodied as well as locatable spatial practices. However, seascape epistemology slides beyond this relationship due to the fluidity inherent in the knowledge and practices. The way I define and use 'āina is fluid, as is the way I define and use the body. This concept of movement and change, even of places and bodies, is at the core of seascape epistemology. I am expanding the notion of seascape into a methodology about the movement of theories, realities, and identities, offering Kanaka Maoli a means of finding specific routes through the seascape, and toward empowerment.

Hau'ofa notes, "Despite the sheer magnitude of the oceans, we are among a minute proportion of Earth's total human population which can truly be

referred to as 'oceanic peoples.' All our cultures have been shaped in fundamental ways by the adaptive interactions between our people and the sea
that surrounds our island communities" (Hauʻofa 2005, 37). The oceanic
metaphor is one of mental movement and travel within a constantly fluctuating world, the seascape as well as the physical place of the Pacific Ocean.
The symbol of water offers flexibility as well as mobility as new routes are
sailed within an "ocean" of possibility.

He'e nalu, as well as hoʻokele and lawaiʻa, become not merely practices
but critical ways of knowing and doing. The practice of heʻe nalu creates a
counter politics to the colonial narrative that has determined Hawaiʻi to
be a fixed geographic land with the sea as a mere boundary, and Hawaiian
identity to be already established and stagnant. Kanaka scholar Rona Tamiko Halualani articulates this sociohistorical production of the Hawaiian
identity as a soft primitivism through a mapping of the islands and the sea:

> Temporal space and geographic representation in maps "naturally
> place" foreigners in Hawaiʻi, as Hawaiian identity is rehauled. A heav-
> ily impacting subject position of Hawaiianness as "naturally placed"
> is created through signifying and representational processes within
> maps that Denis Wood deems as the "culturalization of the natural"
> and the "naturalization of the cultural." Hawaiians, thus, are to be un-
> derstood through the "natural" elements of what is already out there,
> which are themselves sociopolitical constructions; through the kind,
> calming oceans, the pleasant tradewinds and breezes, and the abun-
> dance, the lushness, of food and land (they are indeed inherently
> calm, pleasant, and rich in generosity of what they have). By iconically
> inscribing the "natural" and the "geographic" via maps and charts,
> Western imperialism imagined and brought into being national mod-
> ern space. The cultural and political production of geography serves
> then to naturalize the colonial occupation and newly established na-
> tivism of Hawaiʻi and its people by the British and later U.S. forces.
> (Halualani 2002, 7)

I ponder the seascape in this work because of its potential and power not
only within Hawaiian epistemology and ontology but also against colonial
structures. Hauʻofa eloquently articulated how Pacific Islanders need to
awaken themselves to "the ocean in us," to the ever-present ocean in our
souls as a powerful potential in the face of such colonial narratives. He sug-
gests, "An identity that is grounded on something so vast as the sea is, should

exercise our minds and rekindle in us the spirit that sent our ancestors to explore the oceanic unknown and make it their home, our home" (Hau'ofa 2005, 33). Hau'ofa is expressing the critical role that ka moana plays, and has always played, in Pacific Islander life, constantly moving and shaping our bodies, minds, and societies. Thus, the links between identity and place, both which are critical to the "indigenous," are not static. For Kānaka Maoli, this politics, these moments and interactions, revolve around a specific interaction with the 'āina rather than with a geocentric model that engages in a proprietary contract with land and ocean.

Seascape epistemology builds upon these concepts and provides a decolonizing methodology for Kanaka by revealing hidden linkages between water and land that speak to indigenous ways of knowing and being, to historical means of political, social, and cultural survival. Seascape epistemology engages a discourse about place that recognizes the ocean's transient and dynamic composition; waves are constantly formed and broken, sucked up from the very body that gave it life. No part of this liquid body is ever stable. Yet something does endure within this space and time: relationships that draw together the sea's collective components through an engagement such as he'e nalu. Seascape epistemology is movement's sound, its taste and color, and it is the fluctuation of a process that joins the world together. The epistemology imagines a world where, as Paul Carter, academic and author, articulates, "the laws governing relationships count, and where the value of passages is recognized" (Carter 2009, 6).

The ocean becomes a metaphor for global unity, pulling together and sustaining humankind. Hau'ofa wrote that he was profoundly struck by a piece he read by Sylvia Earle in the October 28, 1996, issue of *Time* that magnified the power and significance of the ocean for the world: "The sea shapes the character of this planet, governs weather and climate, stabilizes moisture that falls back on the land, replenishing Earth's fresh water to rivers, lakes, streams—and us. Every breath we take is possible because of the life-filled life-giving sea; oxygen is generated there, carbon dioxide absorbed. . . . Rain forests and other terrestrial systems are important too, of course, but without the living ocean there would be no life on land. Most of Earth's living space, the biosphere, is ocean—about 97%. And not so coincidentally, 97% of Earth's water is ocean." (Hau'ofa 2008, 52). To speak of a seascape epistemology then, is to address global as well as local issues in relation to not only identity, politics, and economics but also morality and humanity. In the face of modernization, the ocean becomes an increasingly critical place to address in terms of regional political colonization as well as global ecological denigration.

The March 11, 2011, tsunami that hit Japan and the April 20, 2010, oil spill in the Gulf of Mexico provide compelling examples of the powerful and sometimes violent relationships between people and the ocean, each profoundly impacting the other. Japan lost 22,000 people within minutes as a forty-five-foot tsunami stormed across its shores. The tsunami was generated by a 9.0 earthquake (one of the largest ever recorded) off Japan's eastern coast, which also created the worst nuclear energy disaster since Chernobyl. According to the *New York Times*, explosions and radioactive gas leaks took place in three reactors at the Fukushima Daiichi Nuclear Power Plant, which suffered partial meltdowns, releasing radioactive material directly into the atmosphere, fresh water sources, and the ocean. Within minutes, the ocean altered Japan's political, economic, and social structures, sucking tons of twisted steel and debris into its westward ocean currents. Conversely, Japanese-made nuclear pollution leaked uninvited into the ocean. As of May 2012, tuna in Japanese waters have been reported to be carrying high levels of radioactivity. The *New York Times* reported on February 20, 2014, that about one hundred tons of highly radioactive water had leaked from one of the tanks at the devastated Fukushima power plant. This provides an illustration of the many mishaps that continue to plague containment and cleanup efforts, as well as the hundreds of tons of contaminated groundwater that still flow into the ocean every day.

Not long before the Fukushima disaster, the 2010 British Petroleum spill inadvertently dumped up to 184 million gallons of oil into the Gulf of Mexico when a drilling rig working on a well exploded a mile below the surface. While this spill highlights the ocean's vulnerability to human destruction—countless birds, fish, deep and shallow coral reefs, seaweeds, marshlands and grasses, and the water molecules themselves were coated with crude—it also shows its power: the oil-coated ocean directly affected the political, economic, and social climate in the Gulf, just as it did in Japan. The interlocked relationship between human beings and the seascape is inescapable. As we continue to deface the sea—dumping waste, polluting runoff, creating greenhouse gases that cause a rise in ocean temperatures that kill coral reef systems and melt polar icecaps and cause acidification— the ocean in turn buries countless swimmers, surfers, voyagers, fishers, divers, and many more under its salty dominion, a domain proven to be both tranquil and tumultuous, nurturing and deadly.

Kānaka Maoli deem the ocean to be the domain of Kanaloa, *ke akua* (the god) of ke kai. Kanaloa brings life as well as destruction, and is revered as both good and bad in Hawaiian *moʻolelo* (oral history). Epitomizing the sea's dynamic character, Kanaloa comes from a foreign land, having migrated from Kahiki with the god Kāne, washing up on the island of Kahoʻolawe (another name for Kanaloa) like an approaching ocean wave and becoming part of the island's genealogy. Kanaloa brought gifts with him; both he and Kāne brought animals such as the pig to Kānaka Maoli, established fish ponds around the islands, and were often known in the back of mountains as water finders: "ʻOi-ana (Let it be seen)!' says Kanaloa; so Kāne thrusts in his staff made of heavy, close-grained kauila wood (*Alphitonia excelsa*) and water gushes forth. They open the fishpond of Kanaloa at Lualuʻilua and possess the water of Kou at Kaupo. . . . They cause sweet waters to flow at Waihee, Kahakuloa, and at Waikane on Lanai, Punakou on Molokai, Kawaihoa on Oahu" (Beckwith 1970, 64).

Kanaka cultural specialist Keoni James Kuoha explains that Kanaloa is also associated with "depths," with deep water, and with "the unseen but present" (Kuoha 2012). Kanaloa's domain holds much philosophical potential that becomes particularly significant within a colonial reality. The power of ke kai vibrates beyond its picturesque paradise image. Its potential oscillates between what capitalistic and state centric images reveal, in what many oceanic literate Kānaka can't necessarily see but can feel: the "unseen but present."

Engaging an epistemology that allows us access to, a relationship with, and skills for constructing a space of political determination within ke kai, is critical for Kānaka ʻŌiwi. The concept of seascape epistemology becomes more than a cultural representation of the ocean through oceanic literacies; it becomes a way of knowing and being through interaction for Kānaka Maoli. The seascape is not merely represented through a specific Hawaiian lens in this work; ke kai comes to involve an epistemology. A seascape epistemology evolves as an interactive and embodied ontology; a kinesthetic engagement and reading of both the physical and metaphysical simultaneously, enabling an alternative epistemology for Kānaka.

For example, when surfing, I have the inherent ability to reflect on knowledge production as a hegemonic language because my oceanic literacy sits outside of dominant literacies, contrasting established structures by displacing them with my body's gestures and defiance of gravity as it glides vertical, diagonal, fast, and smooth. My literacy is not a matter of being "fluent" in the language of standing up on a board and riding a wave. My literacy is valuable as a way of moving through the ocean (and life) by anchoring myself within its fluctuations. This approach to knowing engages ke kai as a historical mechanism for re-imagining identity. Ke kai offers both corporeal decolonization, through physiological gestures, as well as psychological decolonization by helping me to rethink what a contemporary Hawaiian epistemology might entail, and to reassess how knowledge is produced and taught. There are political rhythms of reactivation and deactivation as Hawaiian ontology is coded through the performance of heʻe nalu, or is conversely overcoded by (neo)colonial structures and thought-worlds that have in part reshaped the islands through development.

The indigenous surfer can become an aesthetic subject whose movements in the time and space of the ocean articulates an ontology and epistemology that opposes the commodification entrenched within the American-settler intelligibility. My work contrasts the disparity between a Hawaiian ontological experience of place with the experience of Hawaiʻi as a place of commodity by assembling a Kanaka Maoli ocean narrative on the microlevel (my autoethnographic moment as a Kanaka Maoli surfer), as well as on the macro-level (an ethnographic mode of representation of ocean knowledge in Hawaiʻi gathered through oral histories, texts, poetry, and artwork). What becomes visible is how identity, through kinesthetic involvement with the ocean, can be deconstructed and reconstructed through movements, imaginations, and a merging with place that honor place-based wisdoms and memories in an era of ecological destruction and detachment, as well as (neo)colonial imposition. The sensorimotor pathways that the body creates for itself engage an oceanic literacy, an embodied "reading" and "writing" of a specific oceanic space, which for Kānaka Maoli are affective, philosophical, and spiritual movements of recovery.

Although the insight here may be open to anyone who has ocean-based knowledge, such as, for instance, a lifelong surfer of Scottish American descent residing in Santa Cruz, California, seascape epistemology is a specific concern of indigenous politics because of what it offers native peoples with colonial legacies. Indigenous politics stimulates an autonomous reimagining of diverse ways of existing and defining one's identity when this right has been forcibly interrupted by geographic, cultural, economic, and religious imposition. This new politics and ethics does not exclude non-Kanaka; its purpose is to include Kanaka in a system that has dispossessed us from our native ʻāina, systematically alienating us from our ethos and marginalizing us in terms of health, education, political power, and socioeconomic status. It is the specific historical location and identity of Kānaka Maoli, as we sit within a stagnant landscape of dominant (neo)colonial structures, that provides seascape epistemology as a tool and critical concept for movement around and through imposed systems, toward self-determined constructions of ourselves.

It is also critical to note that in my analysis of specific neocolonial ideologies within the surf tourism industry, those ideologies transported in the luggage of contemporary surf tourists that reinforce specific knowledges, stories, and theories about entitlement and apolitical movement, it should be understood that as a surfer and traveler, I also move inside this system. I have journeyed to Samoa; I frequent surf events and surf establishments; and I consume surf products and images. My discussion doesn't aim to comprehensively and unconditionally condemn the industry, surf tourists, or surf tourism. Not all surfers or corporations travel in the same way. Motion and exchange are natural and potentially positive phenomena.

The element of neocolonialism becomes active in surf tourism when attempts are made to efface a people's history and autonomy for profit, making a people's land and sea an available feast of enjoyment and consumption—not only as something to be desired but as attainable on surfer's terms and conditions. This neocolonial context illuminates what is at stake ('āina and self-determination), and why seascape epistemology, and the ocean-based knowledges within it, are relevant and necessary. The surf tourism industry has established Hawaiian identity and place as something static, to be conquered, controlled, and exploited. Seascape epistemology disrupts that narrative and economy at the levels of both sensation and thought through an embodied reimagining and re-creating.

Theoretical Framework

Working from an intersection of knowledge systems, a paradigm endorsed by Subramani, professor of literature at the University of the South Pacific in Fiji, requires seascape epistemology to engage in meaningful conversations across differences and disciplines so that it can assess divergent claims about knowledge. Ultimately, there must be a mixing of roots and new routes to keep pace with the variable forces of change in the modern world, and to inspire cultural innovation within the fields of indigenous politics and indigenous studies. Kānaka Maoli are accustomed to traveling, and it should be understood that issues of identity, culture, and tradition can take place within the contexts of nationalism, globalization, and diaspora (Diaz and Kauanui 2001). My focus is less on sources of neocolonialism in Hawai'i, and more on how Kānaka Maoli can be "modern," both indigenous and global, while reaffirming autonomous definitions of ourselves.

Shaping a seascape epistemology involves compiling the language by which it is articulated. This compilation forms a sort of archive documenting some of the many ways that the ocean is known and understood by Kānaka Maoli. I term this collection of oceanic literacy an "archive" because it is anchored in history and genealogy yet is a living archive that expands as Kanaka knowledge evolves. This archive always remains relevant through adaptation. My methodology in the creation of this archive requires analysis of historical mo'olelo and *mele* (chants), but also current interviews and ethnographic observation. While the development of seascape epistemology is methodologically dependent upon the genealogy of mo'olelo, it is also dependent upon ocean experience and sensibility. It is not possible, then, for me to articulate this indigenous epistemology by simply reading Hawaiian texts or by reading the genealogies behind the words describing ke kai. Necessary too to truly embrace the literacy within seascape epistemology are the articulations of embodied sensations and contemporary experiences. The goal is to create an epistemology through which specific oceanic literacies can travel into a contemporary world as relevant ways of knowing for Kānaka Maoli.

I have come to realize that the "visual" is extremely important to my work in developing this archive of ocean literacy; the seascape involves a specific way of approaching knowledge that embraces visual interactions with and conceptions of the 'āina. In expressing Kanaka notions of ke kai within an epistemology, I need to be able to "see" the ocean. Ke kai is fluid, and a Kanaka concept of ke kai must therefore be explored through fluid mediums in addition to texts. Archival and contemporary photographs as well as historical and current oral histories provide a visual image of how Kānaka Maoli move(d) on and interact(ed) with the seascape. The contemporary Kanaka poetry included in this book colors the ocean with a modern interpretation and understanding of relationship to ke kai. The images provided by Kanaka artists also help to enunciate the concept of kai from a Hawaiian perspective. I am also very influenced by the art of Native Hawaiian surfers, fishers, navigators, paddlers, divers, hula dancers, musicians, and artisans. Their work also guides the understanding of oceanic literacy as an

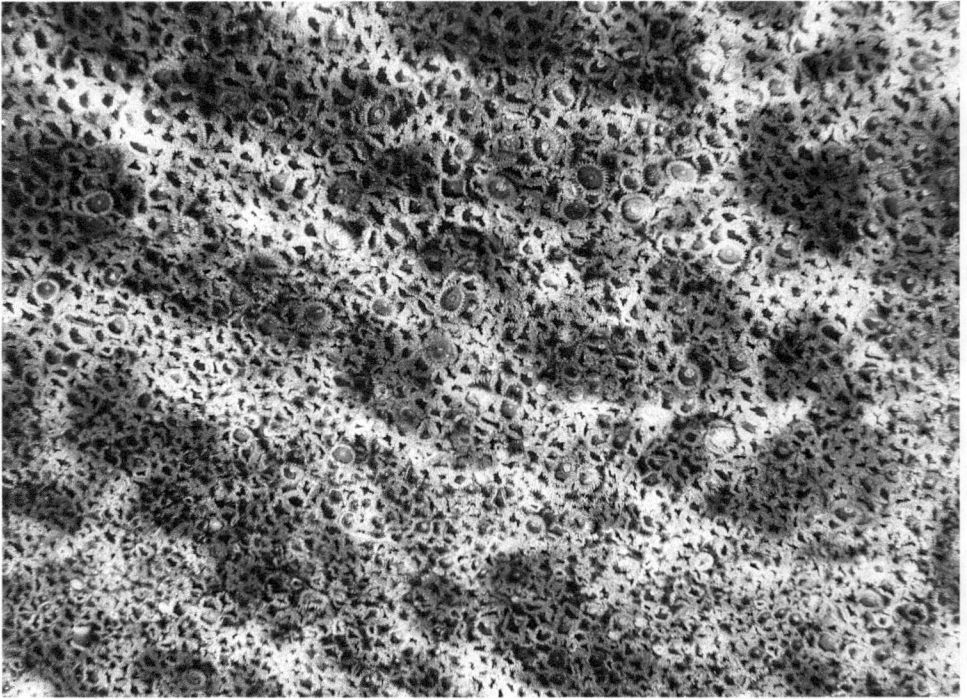

FIGURE I.3. "The Textured Seascape," 2009. Photograph by Russell J. Amimoto.

organic creation, fluidly carrying different Kanaka layers of oceanic knowledges and relationships along a single current of Kanaka waves. Revealed are the sensations of how it feels, smells, and sounds to ride upon the ocean, to (re)discover islands, to hear the fish and he'e (octopus) in the hunt, and to see our genealogical and historical connections to the seascape literally through a Kanaka "lens."

People are not the only entities we engage as indigenous academics conducting research in our contemporary communities; we must also engage ancestors, gods, oceans, rivers, valleys, winds, rains, and stars, which are a part of our communities. A shift in research definition and focus will better support indigenous ways of knowing and being, and thus indigenous self-determination.

Solutions and means of empowerment for Kānaka must be born from an internal origin and strive to function within an indigenous extraction of decolonizing ideologies drawn from Kānaka, enabling me to produce indigenous-based knowledge, to operate from within, and to ground my research strategies in indigenous epistemologies. Grounding a theory or research strategy in indigenous epistemologies is a strong form of decolonization in itself, as is the decision to function within the dominant system, learning how to manipulate it for benefit.

To effectively grasp the notion of seascape from multiple Kanaka perspectives and sources, the tactic for writing this project will be to approach it is as a collage, gluing the diverse and individual seascapes and sources together into one overlapping and blended image of Hawaiian land and sea. The pieces will not lock together like a puzzle but instead overlap and remain independent, incomplete, and infinite. The text will take the form of a paper of poetic literacies, experiential colors, and sets of theories that have shifted outside Western critical paradigms into a reinvented Kanaka concept.

Thus, the key theorists I turn to in this work include Epeli Hauʻofa, with his image of a "sea of islands"; Teresia Teaiwa, as she (re)defines the term "native" in relation to movement and fluidity; Vincente Diaz and Kēhaulani Kauanui, both of whom emphasize the importance of place and being situated "in-between" (Diaz has also done much work on seafaring in Oceania, which I draw on); Rob Wilson, who helps me articulate ke kai as a cultural space; Subramani, as he calls for a regional epistemology based around the ocean; Lilikalā Kameʻeleihiwa, with her perception of Kanaka cosmology, time, and place; Mary Kawena Pukui and Samuel Mānaiakalani Kamakau, with their historical reading and documentation of Kanaka knowledge in *mele* (chants/songs) and moʻolelo, and Noenoe K. Silva, with her interpretations of Kanaka moʻolelo.

I also look to scholars outside indigenous realms to help shape and articulate seascape epistemology. While Kanaka and Western epistemologies are distinguished by fundamental philosophical, cultural, ethical, and geographic origins, they engage through time, space, and place. Indigeneity is both a local and a global interaction. This appears to be a tension but is ultimately a continuous negotiation between roots and routes.[3] My investigation bridges the divide between a European critical philosophy trajectory and an ocean-based indigenous imaginary and set of identity practices. I reinflect the Western philosophical tradition in order to frame the Hawaiian issue of an ocean-based epistemology, translating seascape epistemology into a critical theory idiom. I hold on to both an indigenous imaginary and Western philosophy's trajectory of ontological and epistemological frames. I am touring through theoretical spaces that were once colonial while hopefully creating a new theory about a specific Hawaiian place-based knowledge, a knowledge that is not new in itself but can be used in a contemporary reality of neocolonial institutions such as the surf tourism industry. I engage the space between Western and Kanaka episte-mologies because there is a historical and cultural relationship, even if that relationship is violent.

To develop the theoretical elements within seascape epistemology, I turn to Martin Heidegger's work on being-in-the-world, or being-there, *Dasein*, to articulate how my concept is a temporal epistemology embedded in a metaphysical ontology. Gilles Deleuze and Felix Guattari help me to enun-ciate conceptions of times "in-between" dominant narratives of time, or indigenous times. The theories of Jacques Rancière and Michael J. Shapiro offer political philosophies through which I can articulate how Hawaiians develop independent voices that disturb the status quo within the spaces and times of the sea. Because he'e nalu is an enactment that engages ke kai, it becomes political for Kanaka Maoli, and it becomes useful in exploring the genealogy of he'e nalu. Manuel DeLanda's assemblage theory helped to formulate my own term "ocean-body assemblage," which is discussed in chapter three.

I also turn to environmental authors such as John Muir, Henry David Thoreau, Walt Whitman, and Rachel Carson for lyrical notions of how hu-mans interact with the ocean. I also rely heavily on the work of political theorist Paul Carter, who offers a brilliant perspective on place making as static as opposed to fluid. I also draw upon the metaphorical work of James Clifford and his theories on travel and diaspora.

I use Western thought and philosophy to develop a Kanaka epistemol-ogy in part because contemporary Hawaiian identities are intertwined with a colonial legacy. This does not mean that the colonial must define us, nor does it infer that we are forced to acknowledge or center our work around colonialism as a locus of power. It does mean, however, that Western thought has touched us, and when refocused, it becomes available for and potentially useful in an indigenous framework. This may not be a popular approach to the articulation of a Hawaiian epistemology, nor is it necessary, but given my very colonized background, this is the place from which I write. Knowledge and theory travel, allowing the dominant or colonial phi-losophy to be hijacked by the identities it marginalizes and re-created into something beneficial and empowering. The Western philosophers I include in this work are employed because they help me articulate how a Kanaka identity, as related to the sea, becomes political and ethical in a modern reality.

Knowledge travels; the knowledge within seascape epistemology, although based on concrete skills and aspects of the seascape, as well as an ontological awareness of connectivity, is not fixed or finite. This is its nature: flexibility and change *alongside* nature. Because seascape epistemology is not purely theoretical in nature, this knowledge will take shape only after I articulate not only the philosophy behind an oceanic literacy but the specific applications of it within Kanaka communities and culture. There is a strong political economy within seascape epistemology that involves changing forces of survival and means of livelihood.

Language

The term "epistemology" is fraught with impurities. A word of Greek origin, *The Oxford English Dictionary* defines "epistemology" as "[f. Gr. ἐπιστημο-, combo. form of ἐπιστήμη knowledge + -λογίς discoursing (see -LOGY).] The theory or science of the method or grounds of knowledge" (*The Oxford English Dictionary* 1989, 338). Epistemology is the philosophy of the nature, origin, and scope of knowledge; it is about an approach to knowledge. There are diverse forms of "knowing," one being the possession of knowledge, intelligence, or understanding about, for example, what the ocean is—that is, the body of salt water that covers over 70 percent of the earth's surface. Another form is the "meta" form of knowing, or a way of knowing through an embodied sense of knowledge, as in knowing one's position in the ocean by interpreting surrounding signs. One does not think of "things" abstractly but through an engagement with these things.

I will use the word "epistemology" as an approach to knowledge in the latter, meta form of knowing that encompasses sensations, carrying us beyond deductive and inductive ways of knowing. A meta form of knowing is specific to Oceanic indigenous epistemology, which is interconnected, embracing the surrounding sea from the shore out past the horizon. In this way, oceanic indigenous epistemology is also connected to territory as knowledge and includes the understanding of and interaction with place. Using the term "epistemology" helps incorporate both the Western and Kanaka aspects of ocean-based knowledge: its nature, truths, and justifications, as well as its means of production and skepticisms. The term allows me to articulate how the literacy within the epistemology is not "knowledge" but a way of knowing that can be translated into other contexts.

The form that the politics of representation takes, the categories used, such as that of ocean epistemology, offers an opportunity to be metaconscious of form and intent. Kanaka "knowledge" is not all alike, and there are many diverse ways to interpret the modes of knowledge production. Manulani Aluli Meyer states, "Hawai'i is a vast ocean of diversity given the nuance of environment, foods, gods, gender, age, class, point in history, or political climate" (Meyer 2003, 85). Thus, this work's interpretation of ocean-based knowledge is one interpretation, based on the information I have gathered, and offers one possible means of empowerment for Kānaka Maoli that will help to clarify a specific cultural form of knowledge relevant to epistemology and will not exclude or refute other Kanaka Maoli knowledges.

Articulating a Kanaka epistemology should ideally be done in 'Ōlelo Hawai'i (Hawaiian language), but as I have not yet reconnected with this part of my Kanaka identity, I draw upon a broader definition of language. Two issues related to language must be addressed regarding my work: the expansion of language to include oceanic literacy, and the articulation of this oceanic literacy and seascape epistemology in English. Hawaiian "language" involves not only a spoken or written word but also the genealogy of history that is specifically Hawaiian. I rely on the language of the ocean to articulate the contemporary Kanaka concepts of seascape epistemology, defining "literacy" as reading memories, ideas, and knowledges written in the land and sea. This is not to say that recovering 'Ōlelo Hawai'i is insignificant or unnecessary in the modern world, nor that the translation of oceanic literacy into English does not fall into the usual traps of any translation. On the contrary, an oceanic literacy aids this critical act of recovery, offering another means of accessing a Kanaka epistemology that continually grows and develops alongside a multisited Kanaka identity.

It is the Western perspective that divided knowledge into diverse disciplines, segregating oceanic literacy from other forms of "knowledge." The very definition of literacy has changed through the time and space of history, and through the history of Western colonization, oceanic literacy has been subjugated to Western standards and definitions. Feminist scholar Ramona Fernandez asserts that "literacy discourses are recursive; they circulate in a closed semiotic system that is infected with Enlightenment ideology," and that what we need to strive toward is a completely new understanding of literacy as a complex and constantly evolving skill, embedded in interwoven sets of knowledges, deployed in in-numerable settings, and using existing and yet-to-be-invented technologies" (Fernandez 2001, 7, 9).

Fernandez speaks of writer Jorge Luis Borges as an example of someone who helps expand the definition of literacy; Borges frames reading and writing within the context of memory, imagination, dream, desire, and possession. These are not functional skills alone, Fernandez explains, but they "exist to give humans access to the universe of knowledge, a universe representing the universe of experience" (3). Thus, not only is there a powerful relationship between the written word and movement toward other states of being, but that "reading" can take on other forms: the environment, people, and events can be read.

Fernandez tells us that a postmodern society requires a flexibility of mind that doesn't rely on decoding and calculating skills, but that can travel across a constantly shifting landscape of knowledge in a tumultuous sea. Our "work" is always changing, so how do we know the "correct" way to become literate within our work? In fact, the changes in our work are changes in our literacy. Fernandez says, "What is needed is a completely new understanding of literacy as a complex and constantly evolving skill, embedded in interwoven sets of knowledges, deployed in innumerable settings, and using existing and yet-to-be-invented technologies" (9). This is critical because, as Fernandez explains, "Literacy is consequential. Lives depend on it. Civilizations rise and fall with it and with them their semiotic systems. In the modern world, national policy, personal and collective investment, business prospects are tied to it" (11). Oceanic literacy allows us to open our pasts to our future. "Imagining and literacy are inextricable because it is only through the imagination that we can create other possible selves, . . . Imagining literate selves allows us, whoever we may be, to envision community, nation, and ultimately world. Indeed, imaging literacy is central to the many necessary acts of making ourselves and the world" (11). It allows us to define what literacy actually is, what it's for, and what it offers and enables.

The fact remains, however, that while expanding the definition of literacy helps to include oceanic literacy within dominant forms of literacy, it does not resolve the fact that this work, written and researched in English, has inherent translation challenges. My hope is for a transparent push toward the articulation of a Kanaka epistemology in English.

The politics in using the English terms "postcolonial," "native," and "indigenous" is also an extensively complex issue. I predominantly use the term "indigenous" because of the connotation it has of connection to place—not drawn from a postmodern vocabulary but as Kānaka Maoli might understand and use it. A person or practice is "indigenous" not solely because of a connection to geographic place or cultural space but also because of how these places and spaces are interpreted. I realize that all three terms are potentially problematic, each carrying a colonial frame of reference, but, as noted, it is not necessarily without purpose to place this project within the historical context of colonialism. It is an affirmation and acknowledgment of the past, with the suffering and oppression, which teaches lessons for a condition of future possibility without requiring that the colonial defines the indigenous nor that the indigenous is forced to center the conversation in opposition to the colonial. It is a mixing and a sitting on the edge, as Kēhaulani Kauanui and Vincente Diaz would state. Kauanui and Diaz contend that Pacific Islanders continue a history of production and destruction through both a participation in and resistance to colonialism, patriarchy, militarism, Christianity, nationhood, development, tourism, literacy, athletics, and other forceful modes of modernity and scholarship (Diaz and Kauanui 2001, 316). They advocate a place "in-between" and on the edge of scholarship and the dominant narrative, in which native studies can exist without relinquishing the groundedness of indigenous identity, politics, theory, method, and aesthetics. Ultimately, there must be a mixing of roots and modernity to keep pace with the variable forces of change in the world.

In reference to those indigenous to Hawai'i, I predominantly use the term "Kanaka Maoli." Kanaka scholar Noenoe K. Silva, explains, "This is an old term seen frequently in the nineteenth-century Hawaiian language newspapers. 'Kānaka' means 'person,' and 'maoli' means 'real; true; original; indigenous.' 'Kānaka' by itself also means 'Hawaiian,' especially when used in contrast with 'haole' when meant as 'foreigner' (Kanaka denotes the singular or the category, while kānaka is the plural)" (Silva 2004, 12). I also occasionally use the term Kanaka 'Ōiwi (Bone/indigenous person) in reference to Native Hawaiians.

"Pacific Islander" is also a term used frequently in this work to represent all indigenous people from the regions named Polynesia, Micronesia, and Melanesia. I use the terms "Moana," "Oceania," and "Pacific" when referring to these regions, but I tend to favor the former two. Epeli Hau'ofa asserts that "Pacific" denotes "small areas of land sitting atop submerged reefs or seamounts," while the former, Oceania, "denotes a sea of islands with their inhabitants" (Hau'ofa 1993, 153). Hau'ofa has, however, questioned his own use of the term "Oceania," a term that does not exist, except as a geological fiction. "Oceania" is also a foreign term. Teresia Teaiwa quotes Hauʻofa, "But we (prefer to) use the term 'Oceania' instead of the 'Pacific' because we are not a tame and peaceful people" (Teaiwa 2005, 23). Teaiwa goes on to say that Tongan scholar ʻOkusitino Mahina offers the term "Moana," because it means "sea" in a number of Polynesian languages, and I encourage this term because of the close relationship that Moana has to the Hawaiian language: "moana" means ocean in ʻŌlelo Hawaiʻi (*kai* is translated as "sea" or "area near the sea," and *moana* infers more of the "open ocean" and can also mean "wide" or "spread out") (Pukui and Elbert 1986, 114 and 249).

"Seascape epistemology" indicates language about ke kai and moana, yet it is critical to note that it incorporates knowledge of both the land and the ocean. Kānaka Maoli perceive the ocean as an extension of the land, a perception reflected in the fact that activities that take place on the land always affect the sea, just as oceanic activities have effects on the health of the land. The Hawaiian word for "land" is ʻāina, which translates as "that which feeds," and can also be considered as "origin," "mother," "inspiration," and "environment" (Meyer 2001, 128). Kānaka Maoli had widely populated the islands of Hawaiʻi by 700 AD, dividing their home into *ahupuaʻa* (pie-shaped sectioned land divisions) that usually extended from the mountains out to the sea and comprised a large valley, or several small ones (Charlot 2005). Lilikalā Kameʻeleihiwa writes, "The word ahupuaʻa means 'pig altar' and was named for the stone altars with pig head carvings that marked the boundaries of each ahupuaʻa. Ideally an ahupuaʻa would include within its borders all the materials required for sustenance—timber, thatching, and rope from the mountains, various crops from the uplands, kalo [taro] from the lowlands, and fish from the sea. All members of the society shared access to these life-giving necessities" (Kameʻeleihiwa 1992, 27).

When I refer to Hawaiian "land" or ʻāina, I will be referring to both land and ocean, because although land and sea are distinguished areas, Kānaka Maoli epistemologically perceive them as connected. For instance, *waʻa* means "canoe" in ʻŌlelo Hawaiʻi, and *waa* is also the name for the form of liquid lava that travels like a canoe down the skirt of the volcano, expanding the land that is born up out of the seabed. Exhibited within ʻŌlelo Hawaiʻi is a realm of interconnected possibility that can create new, indigenous terrain, and that helps to mobilize Hawaiian bodies as ever-shifting and negotiating beings.

For Kānaka Maoli, the link between identity and place, both of which are critical to the indigenous, are not static. The politics, the moments and interactions, revolve around a specific interaction with the 'āina rather than a geocentric model that engages in a proprietary contract with land and ocean. It is both a philosophical and physiological, metaphorical and material relationship specific to Hawaiian ontology and epistemology. While a Kanaka epistemology is dependent upon a relationship to 'āina, this relationship is neither absolute nor predefined.

"Oceanic literacy" speaks to the specific ocean-based knowledges of ka 'āina (the land), that are employed within seascape epistemology. The specific literacies of he'e nalu, ho'okele, and lawai'a explored in this book are all living knowledges grown (and growing) from a living archive of Hawaiian mo'olelo, mele (song/chant), oli, performance, and artwork. Ho'okele is the literacy of navigating through the ocean using only the seascape for guidance: the stars, moon, sun, waves, and wind. It is how all Pacific Islanders traveled, traded, migrated, and fished the ocean for centuries. Lawai'a is the general term for fishing in 'Ōlelo Hawai'i, including traditional techniques such as lau nui fishing (with a large net set by canoes) or fishing for octopus with a cowry shell lure. Lawai'a remains a critical literacy for Kānaka Maoli, and for all Pacific Islanders, for cultural and economic subsistence that continues to be engaged through indigenous epistemologies and ontologies.

The ambition of this living archive of sea-based knowledges is to express their theoretical and epistemological significance for contemporary Kanaka Maoli, and to suggest how an archive of oceanic literacy should be approached; how the knowledges of he'e nalu, ho'okele, and lawai'a should be accessed, studied, and experienced. I term this collection of oceanic literacy a "living" archive because, while it is anchored in history and genealogy, it continues to expand and grow as Kanaka knowledge evolves. The archive included here is only a taste; it does not include the greater majority of sources or stories in Hawaiian culture related to ke kai. What this sample strives to stimulate is a discussion about the significance of collecting a living archive of oceanic literacy that contributes to Subramani's call to excavate a body of Oceanic knowledge for "Oceania's Library" with the aim of articulating a regional epistemology. I suggest that this living archive must be read like the ocean, as an organic and ever-changing body of perceptions, nuances, and kinesthetic movements.

A critical portion of the language included in the living archive is mo'olelo. The mo'olelo included here are oral histories from the memories of those who wrote them down in the Hawaiian newspapers at the turn of the nineteenth century (Nāmakaokeahi 2004).[4] Silva writes, "Mo'olelo were sometimes said to have been translated from the oral tradition, however, it is important to understand that written forms of moʻolelo were authored. That is, each of the authors of the many moʻolelo wrote their own versions, using both mnemonic devices from the oral tradition and literary devices that developed over time. Thus, moʻolelo appeared in very specific historical contexts as creations of authors who were often also political actors" (Silva 2004, 160). For this reason, I have chosen to predominantly use Kanaka sources for moʻolelo (as well as mele and oli) to minimize political agendas of colonization, intentional or not.

The development of seascape epistemology is methodologically dependent upon the genealogy of moʻolelo, but it also requires ocean experience and sensibility. It is not possible for me to articulate this indigenous epistemology by simply reading Hawaiian texts or by reading the genealogies that rest behind the words describing ke kai. I need to be able to "see" the ocean. I must articulate the quiver of senses and experiential requirements necessary to read oceanic literacy and provide the cultural context with which to approach this body of ocean knowledge so that seascape epistemology can be effectuated. Necessary are the articulations of how it feels, smells, and sounds to ride upon the ocean, to (re)discover islands, to hear the fish and heʻe in the hunt, and to see our genealogical and historical connections to the seascape, through an actual Kanaka lens. I am therefore necessarily attentive to the voices of Native Hawaiian surfers, fishers, navigators, paddlers, divers, hula dancers, musicians, and artisans as I attempt to translate what the ʻāina is telling Kānaka Maoli.

Articulating these experiential and embodied knowledges about the sea can't be translated directly into "knowledge" because knowledge is shaped by a discourse of language. "Studying" this knowledge requires multiple mediums of expression, explanation, and depth. A modern indigenous epistemology anchored in a contemporary indigenous interpretation of the seascape requires layers. The interviews and art included in this work help to engage our senses, mimicking as best as possible a reproduction of the knowledge within seascape epistemology. Seeking a contemporary Kanaka epistemology about the seascape requires the uncovering of a feeling, a visually learned skill, and a relationship to place that is more than physical; it is also emotional.

The shape that the politics of representation takes, and the categories used, such as that of ocean epistemology, offer an opportunity to be meta-conscious of form and intent. Kanaka knowledge is not all alike, and there are many diverse ways to interpret the modes of knowledge production. Thus, this work's interpretation of ocean-based knowledge is one interpretation, based on the information I have gathered, offering one possible means of empowerment for Kānaka Maoli without excluding or refuting other Kanaka knowledges.

The literacies evoked in this work—Kanaka, indigenous, and Oceanic—are interrelated as well as differentiated from one another and from seascape epistemology. Seascape epistemology is a Hawaiian way of knowing and being, but seascape epistemology also draws from other indigenous experiences and theories—from Hauʻofa in Tonga, for instance, or Teaiwa in Fiji. In this way, the specific oceanic knowledges within seascape epistemology reflect the larger conceptual importance throughout all of Oceania regarding the notion of travel on waves. Kiribati poet Teweiariki Teaero writes in his poem "Ocean Heart Beat,"

> These insistent waves
> Tireless travellers
> From another age
> Come foaming
> To the shore
> Smiling endlessly
> Covering many miles
> Over this shimmering
> Blue blood of Oceania
> Beating a beaten path
> To the wary shore
> Keeping perfect time
> To the rhythm
> Of the beat
> Of the heart
> Of the deep
> Deep ocean
> Forever
> (Teaero 2004, 85)

Within Oceania there are diverse and distinct notions of seascape with distinct oceanic literacies. Each notion of "scape," however, containing elements both similar and unique, unfolds into a larger concept of an "oceanic" connection between the mountains, beaches, rain clouds, bays, reefs, waves, birds, moon, and stars of Oceania. A Kanaka oceanic knowledge of surfing is different from a Kiribati oceanic knowledge of surfing, as the physical geographic differences between the two island nations produce different types of waves and thus different ways of riding them. Kiribati fishers might surf their fishing canoes on outer reefs, while Kanaka surfers ride their boards close to the sandy shores. The specific knowledges and techniques used vary, but both share the enactment of surfing, physically and metaphorically. The oceanic knowledge is significant for both, and both are indigenous knowledges.

These distinctions are important to make as I develop seascape epistemology, because the concept is in part built from theory and thought from the fields of indigenous and Pacific Island studies. While I recognize the similarities, differences, and importance of both the physical and metaphorical oceanic literacies of Kiribati (or of any indigenous people) and Hawaiʻi, application of seascape epistemology, for the purposes of this work, focuses only on Kanaka Maoli.

Chapter Outlines

The first chapter of this book distinguishes between the movements and languages of Kanaka Maoli surfers and those of the surf tourism industry. A Kanaka surfer becomes more than merely a body riding a wave; he or she can also become political through the sensibility of an act that rearticulates a Kanaka way of knowing that includes indigenous history, values, beliefs, and determinations that have been marginalized. Practicing heʻe nalu within the neocolonial reality of the surf tourism industry redistributes what is allowed to be seen and heard by asserting autonomous voices in order to (re)connect. The surf tourism industry in Hawaiʻi becomes a colonial system that effaces indigenous history and place names, and imposes a specific narrative about Hawaiian identity, violating the critical relationships Kānaka Maoli have to ke kai today.

The second chapter develops the specific oceanic literacy within seascape epistemology, articulating the ways in which a surfer, navigator, or fisher sees, smells, hears, tastes, and feels ke kai. Immersing the body in the ocean enables an affective reading of the ocean's rhythms, which speak to political and ethical ways of seeing or hearing because they expand the ways in which one exists. The political and ethical potential within oceanic literacy also emerges through historical and contemporary discourses, place names, stories, and performances in and about Moana, which are included in this chapter. Oceanic literacy presents an alternative way of reading and writing inside places.

Chapter three further develops the concept of seascape epistemology as an embodied and emotional ontology for Kānaka Maoli, which involves an engagement with ke kai in such a way that indigenous identity becomes mobile as the body merges with the fluid ocean. This ocean-body assemblage joins the rhythms of the seascape with the self, enabling a way of moving that is flexible and complex, both affective and intellectual. Seascape epistemology helps to repartition and redistribute dominant systems of knowing and engaging the world for Kānaka Maoli through an indigenous construction of both time and space found between dominant temporal and spatial constructions. Focus shifts onto what cannot be seen through orthodox lenses. Brought back into the foreground, through an ocean-body assemblage, are the white noises of the wind billowing through the clouds and shimmering across the sea's skin, sounds and sights normally drowned out by the call of capitalist and political agendas. Seascape epistemology is about knowing through movements of the body situated within places—movements that have the potential to shape and to (re)create the places we inhabit.

The fourth chapter sails into the specific oceanic literacy of hoʻokele to better articulate how distinct ways of knowing the world construct specific identities as related to our surroundings. How we read the seascape influences how we move through it, constructing distinct ideologies that affect our realities and relationships with the surrounding world. I discuss how historical European ways of traveling on the seascape carried an ideology that distinguished land from sea so that entering the ocean was to enter a mysterious place "out there." The sea, and all that was encountered in it, needed to be controlled in order to "get across." Established was a duality between "us" and "them," between the "civilized" and "wild" worlds. In contrast, Kānaka Maoli have always perceived ke kai as a place of genealogical significance, and thus travel across it never took them far from their own being. Moana was not to be controlled but connected to. The knowledge within hoʻokele illuminates a Kanaka epistemology about movement that draws the world together, fostering an ocean-body assemblage that honors our human relationships and responsibilities to each other and to the places we voyage through.

Chapter five gives concluding thoughts about how the oceanic literacy within seascape epistemology can be applied in *ka hālau o ke kai*, an ocean-based education and community center, where place-based and practice-based education is emphasized by putting youth into the time and space of ke kai, allowing them to touch, see, smell, and taste the seascape.

This work aims to invigorate our imaginations to (re)integrate the ocean back into our epistemological and ontological views—a vital source of survival, movement, history, and genealogy for Kānaka Maoli that has become an internationally commercialized symbol of recreation, "lost" paradise, and consumption by the tourism industry, mass media, the U.S. military, and American politicians. I approach this project with humility, acknowledging the very profound historical and cultural depths within a Kanaka conception of ke kai. The goal of this book is to not necessarily, or not only, critique dominant ideology and politics but also, in the process (after a strong critique is established or provided as a foundation, because this process is never "finished"), to open up new "spaces" and "places" for Kānaka Maoli to expand, and to resist (after contesting) imposed systems, identities, and self-definitions. This book strives to articulate how the ocean helps us to re-create, reaffirm, and return to conceptions of knowing that are plural and progressive by interacting with a space and place that holds so much significance for so many of us, on so many levels.

It is my aim in this work to revitalize not only my own Kanaka heritage, to which I have clung through my connection with ke kai, the sea, but also to rearticulate the indigenous-based oceanic knowledge critical to Kānaka Maoli as an epistemology that allows for a break from the established, idealized, and marginalized identities in today's modern world. I also articulate why this regional literacy is valid for Kānaka Maoli today as an important tool in the struggle for self-determination—how oceanic literacy can offer a new, alternative way of approaching the relationship between knowledge and power. I explain what makes oceanic literacy empowering for Kānaka, and how both the applied and conceptual or aesthetic aspects of this literacy are modified or transformed in contemporary Hawai'i.

Ocean-based knowledge is not a new knowledge for Kānaka Maoli; it has been a focal point of Hawaiian culture and life since Kanaka ancestors sailed to Hawai'i two thousand years ago. Kānaka Maoli have always passed on and practiced oceanic literacy. Fishers know the tides and the seasonal patterns of the marine life and how to sustainably interact with it. Surfers harness the power of the waves; they know the reefs and respect the life, recreation, and health they give. Sailors and navigators know how to use the ocean for transportation and as a directory. And canoe builders craft vessels that do not challenge but yield to the power of the ocean so as to function harmoniously within it. What I propose in this work is a new application of this traditional, indigenous knowledge within academia, as well as in the community today. Seascape epistemology is a knowledge and a literacy, which validates the Kanaka voice in academia and speaks of alternative ways of reimagining politics and ethics. This indigenous oceanic-based knowledge provides an indigenous perspective from which to view the potential for travel and discovery, for movement above and between power structures. It provides an indigenous perspective of thinking, being, and knowing through the seascape, which challenges the dominant perspective of a static "landscape."

Methodologies for applying seascape epistemology within the surf tourism industry lie in the larger goals of education of, and participation and leadership by, Kānaka Maoli in contemporary society. A discussion of the application of seascape epistemology is engaged in the final chapter of this book, but the primary goal of this project is to establish an alternative epistemology to place the shores and depths of the ocean in a new context for Kānaka Maoli from which point we can continue to explore means of self-empowerment.

HEʻE NALU Reclaiming Ke Kai

Sliding into Remembrance

Afloat in the same nutrient-rich water that has been circulating on Earth, in various forms, for three billion years (Farber 1994), Kanaka surfers tap into the wisdom of how waves move. *Na Nalu* (waves) have arms, fingers, and legs— many legs—a face, a back, and lips. Waves move with a gut pushed by forces from the universe: the moon's currents, the sun's reflecting rays, the pull of gravity, and rotation of the earth. Yet each wave finds individual expression, leaping off reefs and crumbling, its lips wet with white saliva. Na Nalu move in general rhythms with improvised expressions of life, blurring some lines and shapes, and coloring in others, like textured tapestries across the globe. Kānaka Maoli have named, studied, ridden, lived off, and lived with nalu for centuries. This oceanic literacy of heʻe nalu enables Kanaka surfers to harness (and sometimes perish to) the power of waves, strengthening their profound ontological connection to the larger sea.

The indigenous Hawaiian oceanic literacy of heʻe nalu, however, now sits within a new struggle over geography in Hawaiʻi, which, as Edward Said explains, "is not only about soldiers and cannons, but also about ideas, about forms, about images, and imaginings" (Said 1993, 7). Heʻe nalu, which has always been an act of performance, establishing and confirming Kanaka social and political order, has in part washed up onto the rocks of a foreign system of control. Ocean "territory" and access to resources within ke kai are the new battlegrounds of surf colonization, a geographic and economic colonization that impairs land and ocean literacy, and access to, relationships with, and an ability to *mālama i ka 'āina* (care for the land).

As many as 7.4 million visitors descend upon Hawai'i each year, 51 percent of whom flood O'ahu's North Shore in search of the "Hawaiian surfing experience." During the winter months, the North Shore coastline population of eighteen thousand triples.[1] During summer months, large sets of these tourists roll into the South Shore (and increasingly the West and East Shores), all on the hunt for the ideal surf adventure.[2] This growing vogue (which began in the 1920s when Duke Kahanamoku and the rest of the Waikīkī Beachboys formally introduced surfing to the world) has evolved into a Western neocolonial presence in Hawai'i, as the contemporary surf tourism industry approaches the Kanaka Maoli seascape as a landscape to be conquered and reaped for entertainment, escape, and profit.

The surf tourism industry is an institution physically swelling in popularity and economic opportunity, having exploded in the last thirty years into a multibillion-dollar niche market that incorporates lucrative vacation packages, instruction camps and tours, equipment sales, mass media, advertisements, and real estate opportunities. Hawai'i has unanimously been designated as *the* pleasure zone for this market. The surf narrative dictates that any hardcore surfer or surf enthusiast must experience the ideal and challenging waves of Hawai'i, and the tourism narrative adds that it can all be done in the beautifully temperate waters of this native, yet safe and accessible, paradise. Together, the surf and tourism industries have created the surf tourism craze. A popular online surf website, Surfline.com, proclaims, "Oahu is a place most surfers will want to visit at least once. . . . it's Ground Central for the modern day sport of wave-riding—birthplace of Duke Kahanamoku, tester of champions, and site of what still ranks as the most high-impact stretch of surfing coast on this earth." For Hawai'i, such a designation means expanded coastal highways, traffic, pollution, the overwithdrawal of water supplies, and the proliferation of oceanfront hotels, private homes, restaurants, surf shops, surf schools, and tours in an overcrowded and polluted ocean.

How can he'e nalu, an indigenous activity rich in political, social, and spiritual significance for Kānaka Maoli, become a neocolonial presence in

Hawai'i? It is my argument that surf tourism has transformed surfing, within the tourism industry, into an international "sport" and capitalistic enterprise prided on discovering, conquering, and experiencing a Western-contrived surf utopia. Surfers have evolved into a breed that overwhelmingly over-looks a conscious awareness of their impacts on the people, oceans, and lands encountered on their surf itineraries. In speaking about the explosion of the adventure-oriented tourists, surf journalist Steve Barilotti quotes anthropologist John McCarthy, "Ironically, these refugees from modernity carry their disease [of escapism] with them" (Barilotti 2002, 92). This is true of surf tourists engaging in an industry that draws upon the sport of surfing but roots itself more deeply in the larger narrative of escape and paradise. The industry has brilliantly tapped into people's natural desire for a lost Eden by leaning on a social, political, and spiritual Hawaiian activity and transforming it into a sporting mission to demand an experience of the exotic, the "frontier," and the authentic.

The "remote" Hawaiian waves advertised by the industry have become legend in the surfer psyche, and every surfer is encouraged to dream of trav-eling to these idyllic waves to become a member of the prided surf clique and fulfill the designated surf fairytale. As more and more surf tourists drop into Hawaiian waves, ke kai becomes prone to this neocolonial fantasy of dominating waves and taking what severs them from the 'āina. Under-scored are the colonial lines etched into the Hawaiian Islands by the Euro-American ideology and geopolitical structure of sovereignty, prioritizing an organization of place around the strong military and capitalist presence in Hawai'i: airstrips across the reef, warships strategically placed along the coast, paved walkways, mass construction of beachfront hotels and shops, private hotel beach zoning, and surf lessons. Surf tourism promotes a static approach to place that cuts and divides ka 'āina for ease of tourist use. For-gotten in this ideology is the body's relationship to place, human memo-ries stored over centuries in the reef and along the coast, memories that act as adherent forces bringing together people and places, arranging these historical memories into contemporary contexts.

The sea itself, however, is the focal point of colonization by the surf tour-ism industry that ideologically establishes ke kai as a place of conquest and domination. The ocean has become a place of encounter and contrast in which the Kanaka Maoli experience of place conflicts with Hawai'i's expe-rience as a tourist commodity. Said has termed colonization as the loss of locality to the outsider. The oceanic literacy of he'e nalu, however, helps

Kānaka recover what has been lost. The process of decolonization or of postcolonial development and politics becomes a (re)establishment of indigenous geographical identity, which is not an "authentic" reality in need of uncovering (such an assumption would support the tourist narratives of an exotic and primitive native and her land), but a (re)vitalizing of indigenous geographic identity through knowledges of and relationships with place.

Heʻe nalu becomes an oceanic literacy for Kānaka that involves a process of (re)creation through both historical memories of, and modern engagements with, the seascape. Heʻe nalu is the cultural enactment that connects the history of Hawaiʻi: Kanaka ancestors surfed the waves of the open ocean from Kahiki, riding Ka nalu to the shores of Hawaiʻi. Every voyage, long or short, culminates with surfing. As an oceanic literacy, heʻe nalu, whether on a surfboard, in a waʻa, with the body, or theoretically in the mind, connects Kānaka ʻŌiwi, both physically and conceptually, to seascape epistemology. Heʻe nalu is an enactment that engages a profoundly nonlinear conception of the environment and of human relationships to it, privileging embodied connections that help to realize multiple and complex constructions of a multisited identity that resonates within the "language" of heʻe nalu.

This language of surfing that speaks to the potential within heʻe nalu begins with the word itself. *Heʻe nalu* (to ride a surfboard, surfing, surf rider, and, literally, "wave sliding") is a word rich with *kaona* (inner meaning), illuminating its profound potential within a Kanaka epistemology. Kanaka waterman and navigator, Bruce Blankenfeld, explains, "Hawaiians are masters at using proverbs and poetic types of things and talking in kaona, you know, not direct. The essence of that is that there's a deeper meaning into what they're saying, there's a hidden meaning, and that's what you have to find" (Blankenfeld 2008). For instance, Mary Kawena Pukui and Samuel H. Elbert's *Hawaiian Dictionary* defines *heʻe* as

1. n. Octopus (*Polypus* sp.), commonly known as squid. *Heʻe mahola*, octopus given for sickness caused by sorcery, as octopus (*heʻe*) would cause the sickness to flee (*heʻe*) or spread out (*mahola*). (PPN *feke*.)

2. vi. To slide, surf, slip, flee (Kin. 14.10). Cf. *heʻe nalu, pūheʻe*. See ex., *puʻe one. ʻO ka mea i hilinaʻi aku iā ia, ʻaʻole ia e heʻe* (Isa. 28.16), he that believed did not make haste. *hoʻo.heʻe*, to cause to slip, slide, flee; to put to flight, rout. *Hoʻoheʻe kī*, ti leaf sliding. (PPN *seke*.)

3. vi. To melt, flow, drip, soften; to skim, as milk. Cf. *heʻeheʻe, heheʻe.*

4. vi. To hang down, as fruit; to sag; to bear breadfruit. See ex., ule. *Laho heʻe,* hernia rupture. (Probably PNP *seke.*)

5. n. Line that supports the mast, stay.

Nalu is defined as

1. nvi. Wave, surf; full of waves; to form waves; wavy, as wood grain. *Ke nalu nei ka moana,* the ocean is full of waves. hoʻo.nalu, to form waves. (PPN *ngalu*).

2. vt. To ponder, meditate, reflect, mull over, speculate. Cf. Eset. 6.6. *Nalu wale ihola nō ʻo Keawenui-a-ʻUmi i ka hope o kēia keiki* (For. 4:261), Keawenui-a-ʻUmi pondered about the fate of this child. (PPN *na(a)nunga*).

3. n. Amnion, amniotic fluid. (PPN, PCP *lanu*)

As I noted in the introduction, *heʻe* can mean "to slide," and *nalu,* which can mean "to ponder," informs how heʻe nalu is an act of sliding into a ponderous state of thinking and theorizing about the world through a Hawaiian context. *Heʻe* also means "to put to flight," and *nalu* also means "to form waves"; heʻe nalu is also the act of putting to flight the formation of waves. Heʻe nalu is not merely sliding across waves, or into a ponderous state, it is an act that helps the seascape to move, putting waves (concepts, spirituality, bodies) into flight.

Sydney Iaukea, a Kanaka surfer and PhD, shares her thoughts on heʻe nalu: "When I think about the word, it makes me reflect on all the sites dedicated to surfing—especially on the Big Island—and the importance that surfing must have held, not only for the physical act itself but for the entire being, for reflection (another meaning of *nalu*). So surfing transcends the physical act and is important for mental well-being" (Iaukea 2008).

Hina Kneubuhl, former instructor of ʻŌlelo Hawaiʻi at the University of Hawaiʻi, offers her interpretation of *heʻe nalu*: "I believe that it might be a metaphor for making love. I know that canoe paddling definitely is (ʻŌlelo Noʻeau has some good sayings about that), and surfing is similar. . . . I think it is safe to say that it was a common practice among chiefs and a great way to impress someone back in the day . . . flexing their skills in hopes of gaining the attention of someone else" (Kneubuhl 2008).

He'e nalu becomes a powerful "language" for Kānaka Maoli within this epistemological capacity, and it also offers a specific ontological potential for (re)connection between Kanaka Maoli and ke kai. To address this connection, which is distinct from the connection between non–Kānaka Maoli and the ocean, I present the narratives of two surfers who have both embraced the ocean-based knowledge, or "oceanic literacy," of surfing as accomplished and lifelong "ocean people." Due to their individual genealogical and cultural contexts, however, these two surfers have distinct ontological and epistemological experiences in and representations of the sea. The two narratives will help to articulate why a Kanaka way of knowing is distinctly "Hawaiian," and thus how it offers Kānaka Maoli political empowerment. This is not an effort to reenforce or create binaries between "native" and "nonnative." As I address in this work, identities are complex and overlapping. Instead, the aim of this work is to distinguish the ontological connection, and thus epistemological differentiation, that makes seascape epistemology a specifically Kanaka epistemology.

The first surfer is Samantha, a fourth-generation American Irish and German woman who grew up surfing the surf breaks down the street from her house and is well respected in the lineups among the most ruthless and aggressive local surfers. While she has proven herself as a big-wave surfer, is sponsored by top surf companies, and travels around the world to experience a variety of waves that have boosted and broadened her ocean-based knowledge, her connection to the ocean was acquired through a specific experience and learned literacy situated in a specifically American historical context. Samantha has mastered the literacy of surfing; she knows how to read the waves, the shifting sand formations, and the circulating winds, and can navigate flowing kelp beds. She has a very keen intuition of the ocean's moods, and has incorporated the act of surfing into her sense of spirituality. For her, surfing forms her identity as defined by her career, her form of economic survival, her social structures, spirituality, and means of physical exercise. Like all American surfers, Samantha sees surfing as a "way of life." Surfing is an act of profound significance with great potential for Samantha, as it helps her develop an intimate relationship with the ocean and shapes the ongoing development of her identity through her specific culture.

The second surfer is Kula, a Kanaka Maoli woman born and raised on O'ahu. For Kula, he'e nalu impacts the evolution of her identity in the same way that it does for Samantha, but as a Kanaka Maoli, Kula is situated within a distinct historical and cultural context. Her ontology is shaped by the coral in the ocean, and her sense of being is connected to this time and space. This alters Kula's relationship to ke kai; she knows the significance ka moana plays in the construction of her Hawaiian identity. She is born from the seabed. Ke kai is where she came from; it is a part of her genealogy. The significance of the shifts in location, culture, and historical identity mean that when Kānaka embrace seascape epistemology from this place of embodied and psychological connection to the sea, the relationship is unique to them. Samantha, despite her profound relationship to the sea, does not have this ontological connection. Samantha's historical and ancestral perspective of the ocean is from a place of Otherness, as separate from the self.[3]

Furthermore, the colonial legacy in Hawaiʻi posits a disparate context in which oceanic literacy and seascape epistemology is applied by Kula because of how this literacy and epistemology are interpreted and used for decolonization. Kula's historical and cultural experience of colonization requires a re-creation, and heʻe nalu is one way Kula can rebuild and reestablish ways of knowing and being that have been effaced. The enactment of heʻe nalu is one way contemporary Kānaka Maoli such as Kula can reaffirm an autonomous identity within a neocolonial reality. Heʻe nalu, when applied through seascape epistemology, can be interpreted and used for a specific political and ethical movement for Kula.

The oceanic literacy of surfing can be embraced by both Samantha and Kula, as can seascape epistemology. What makes both oceanic literacy and seascape epistemology "Hawaiian" when applied by Kula, and by all Kānaka, is the historical and cultural context found in the common Kanaka language. The "language" of which I speak is the language of the common Kanaka history and the common cultural context found in the genealogies of our moʻolelo and mele, our oral history. This language determines a Kanaka ontological connection to the sea and provides the framework for a specifically Kanaka epistemology. It is a language specific to Kānaka Maoli because it is our history and genealogy. All Kānaka, even without physical access to or experience with the ocean, can still embrace the philosophical potential of an epistemology based in this ocean knowledge because all Kānaka share the "language." On the contrary, non-Kānaka who are oceanic literate, such as Samantha is in surfing, do not obtain the same epistemology because they do not have the language: the genealogy of history, the ontological connection that arises from this history, or the historical context of being on the receiving end of a colonial legacy, all of which offer Kānaka ʻŌiwi a particular potential for political empowerment. The applied knowledge is available to both Samantha and Kula, but the theoretical significance and consequence of the epistemology, as I use it, is Kanaka because it is based in a language that is specifically Hawaiian. This epistemology is critical to develop today because of the significant colonial legacy that Hawaiian knowledge, language, and ʻāina has and continues to have.

The Development of Surf Tourism in Hawaiʻi

Throughout the 1820s, heʻe nalu was still a widespread activity for Kānaka Maoli. As the century progressed, however, the condemning voice of the missionaries rang louder and stronger, denouncing Kanaka practices, including surfing, as sacrilegious and hedonistic. Frolicking in the water half naked was seen as not only irresponsible but culturally inappropriate to the missionaries' ambitions. Simultaneously, hundreds of thousands of Kānaka Maoli were perishing due to the introduction of Western diseases, putting not only the Hawaiian population but also the culture in peril.

Anthropologist Ben Finney and author James Houston write, "As far as sport and games were concerned, the most immediate effect of the 1819 revolution was the lapse of the annual Makahiki festival ultimately tied to the god Lono. . . . The Makahiki's lusty stimulus had been of prime importance in keeping sports and games alive and fresh and in maintaining public support" (Finney and Houston 1996, 53). The systematic deconstruction of Native Hawaiian society, religion, and political systems that resulted from the overthrow of the monarchy affected Hawaiian activities that both relied upon and preserved the culture. Finney and Houston continue, "For surfing, the abolition of the traditional religion signaled the end of its sacred aspects. With surf chants, board construction rites, sports gods, and other sacred elements removed, the once ornate sport of surfing was stripped of much of its cultural plumage" (53). Hiram Bingham, chief American missionary, defended missionary policy, and denied the accusation that churchmen suppressed Kanaka pastimes. Regarding surfing, he said, "The decline and discontinuance of the use of the surfboard, as civilization advances, may be accounted for by the increase in modernity, industry and religion, without supposing, as some have affected to believe, that missionaries caused oppressive enactments against it" (54).

In fact, however, missionaries did instigate oppressive enactments against Kanaka practices, including cultural and political sovereignty. Their ambitions reached beyond the church and into every aspect of Kanaka society and life. It was the event of religious conversion that initiated the demise of Kanaka polity beginning in the 1840s. Juri Mykkänen, assistant professor of political science at the University of Helsinki, states, "As a collective, the mission tried to dispel the deeply rooted conviction, held up by several other conspicuous foreigners, that the mission had more than mere spiritual interests in altering the traditional course of Hawaiian chiefship. Indeed, the missionaries had been accused of doing what they resolutely denied: interfering in Hawaiian politics by inducing the chiefs to establish puritanical regulations, and thereby turning themselves into effective rulers of the islands" (Mykkänen 2003, 10).

Mykkänen argues that because Native Hawaiian understanding of ruling involved a matter of mediation between the worldly and the divine, the mission's involvement in Kanaka politics was a natural step toward their goal of complete conversion. Missionary thinking, he states, was imperialistic, based on a "biblical version of reality that could be applied to almost anything, giving rise to gross occasional misapprehensions and frequent conflicts with those foreigners, mainly their fellow Americans, who did not share the demanding tenets of this universe" (18). What began was a multilayered and complex colonization of Kānaka Maoli, an ambition supported and eventually taken over by businessmen in the islands during the time of the overthrow and annexation. To truly colonize this new territory after 1898, however, Hawai'i needed to appear desirable to not only those already in the islands but to Americans on the mainland as well. The tourism industry came into play, capitalizing on the once immoral activity of he'e nalu, which was suddenly romanticized and became an adventurous selling point for the new territory.

Colonial political and economic ambitions in Hawai'i led to a specific cultural production of the islands fabricated by and for the West so that tourists could experience an idealized, exotic, and primitive place that was also beautiful and welcoming. Christina Bacchilega, professor at University of Hawai'i, argues that photography valorized this image on paper, offering a "visual vocabulary" that endorsed the economic and ideological ambitions of the businessmen in the islands. In this way, tourism served Western interests at a crucial political juncture in Kanaka history. Bacchilega explains how a specific cultural production emerged in local literature based not on Kanaka history or culture but on the imagined narrative of the West: "Often accompanied by photographic illustrations, these English-language narratives were not for Native Hawaiians, or even Hawai'i residents, but primarily for people with little knowledge of Hawai'i—prospective tourists and settlers. Americans who wanted to know more about 'their' new Territory, and whose support for the Americanization of the islands had to be won" (Bacchilega 2007, 62).

Bacchilega continues with examples of such publications and photographs from one such English-language tourist magazine, *Paradise of the Pacific*, which was devoted to "the business interests of the Hawaiian Islands" that surfaced in 1888. Edited by Thomas G. Thurm, this magazine declared in 1893, the year of the overthrow, that due to its devotion to the interests of the islands, it was "most positively in favor of annexation." To successfully annex the islands, American businessmen such as Thurm needed to present them as a valuable and desirable place. Bacchilega writes, "*Paradise of the Pacific* variously but consistently promised that the 'stranger in Hawaii is sure to meet with a cordial welcome' and to enjoy a certain standard of modernity and comfort; however, to be distinctively attractive, Hawai'i's image also had to celebrate 'its picturesque isle[s].' Articles and illustrations recording journeys to the Mauna Kea volcano, the majestic Nu'uanu Pali (*pali*, cliff), or the 'uncultivated gardens' on Moloka'i supplied the necessary exotic touch to make Hawai'i a unique tourist destination, still supported by a 'soft primitivism'" (68).

Surfing played a significant role in this purpose. The "visual vocabulary" of picture postcards first took effect in 1898, and surfing was a strong voice within this new visual language as a sellable subject. The Moana Hotel opened in 1901, and stretching from the beach fronting the hotel into the ocean was the Moana Pier, an ideal and popular location from which photographers and filmmakers captured numerous images of surfers (Brown 2006).[4] The images of surfing captured on film (many of the most popular being taken from this pier) helped to lay the foundation for the creation of a promotional Hawai'i: an idealized, tropical ocean fantasy. The making of Hawai'i into a tourist destination, an open space available for exploitation and entertainment, was supported by the appropriation and manipulation of he'e nalu through advertisement.

For example, postcards showed young women posing with surfboards, attracting tourists to what was being portrayed as both a modern *and* an exotic place. In the caption for the image shown in figure 1.1, DeSoto Brown, Kanaka archivist at the Bishop Museum, writes, "Perky young Gertrude McQueen shows off fashionable beachwear on Waikīkī between 1915 and

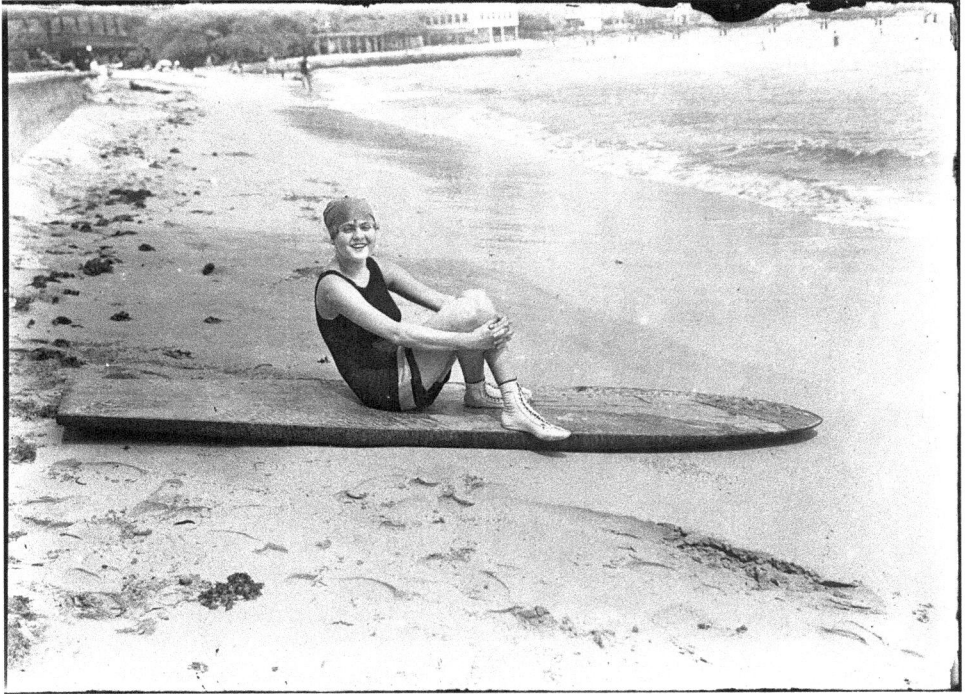

FIGURE 1.1. Postcard of young model on surfboard, in Brown (2006, 40).
Courtesy of the Bernice Panahi Bishop Museum.

1920, including lace-up boots." Yet he notes that, considering the weight of
the board she sat upon, this young model "probably was not able to make
much use of it—or even lift it" (40).

By the 1940s and '50s, much of Hawai'i's image included Waikīkī Beach,
and surfboards became unmistakenly "Hawaiian." Numerous films and
programs were shot in Waikīkī that featured or used he'e nalu as a thrilling
and intriguing aspect to Hawai'i. Examples include the films *The White
Flower* (1923), *Bird of Paradise* (1932), and *Waikiki Wedding* (1939); the
television program *Happy Hawaii*; and the 1934 travelogue *The Island of
Oahu*. *The Island of Oahu* was one of six travelogues commissioned during
this time by the Hawai'i Tourist Bureau, and was produced by a Hollywood
company, illustrating the influence the tourist industry and American
corporations had in how Hawai'i was portrayed to the world (63). Figure
1.2 shows the title shot of *The Island of Oahu*. Note the subtitle: "Territory
of Hawai'i, U.S.A."

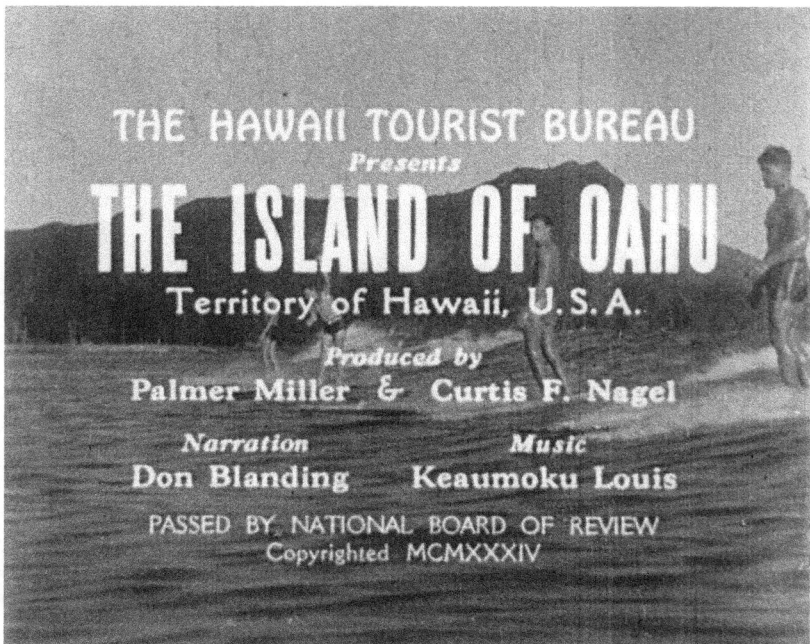

FIGURE 1.2. Title shot of 1934 travelogue "The Island of Oahu," in Brown (2006, 63).
Courtesy of the Bernice Panahi Bishop Museum.

Americans were vying for control over how Hawai'i and Kānaka 'Ōiwi were presented to the mainland, including control over the newly "discovered" activity of surfing. The Western narrative asserts to this day that it was former Chicago newsman Alexander Hume Ford who in 1907 "stepped off a boat in Honolulu and began a crusade to revive the 'royal sport of surfing'" (Timmons 1989, 25). Hume believed that "the white man could learn all the secrets of the Hawaiian [surfer]," and started the Outrigger Canoe Club in 1908 (Walker 2008, 95). The club began as a whites-only surf club where the elite could gather to engage in a paradoxical act of negation and appropriation: they were participating in and excluding the indigenous population from a Kanaka practice that their nation's church had condemned only years earlier.

Although the process of colonization profoundly deterred and inhibited the practice of he'e nalu in Hawai'i, Kānaka Maoli never ceased to surf; they never surrendered the seascape as their place of identity. He'e nalu became a rare but not extinct practice in Hawai'i, with most Kānaka Maoli surfing only a handful of reefs at the turn of the twentieth century, mainly on Oʻahu and Maui. In fact, heʻe nalu not only survived, the surf became a puʻuhonua, a sacred place of refuge, for Kānaka Maoli, as Westerners initially had no systematic interest in or form of control over the ocean (95).

A systematic interest in control over the Hawaiian seascape began to grow, however, as the sport of surfing grew and transformed on the U.S. mainland. Tracing Americans' historical transformation of the sport into a neocolonial industry, it is clear that it was not until the aforementioned Californian surfers of the 1950s that the ambitions of surfers themselves began to change. Surfing traveled. Immediately the space in which surfing existed began to transform, as did its ideology, mode of perception, and tactics of power. The American myth making that reinterrupted surfing was spearheaded by the release of a film in 1959 called *Gidget*. This surf film was adapted from Fred Kohner's 1957 book based on his daughter Kathy's journals telling of her experiences with surfers at "The Point" in Malibu, and it idealized white, middle-class teenagers engrossed in a carefree and sun-filled life on the beach. Although the focus of the film was not necessarily on surfing, it instantly transformed the lifestyle that surfing brought with it into a desirable ideal within the reach of every American adolescent. The number of surfers in California grew from about 5,000 in 1956 to an estimated 150,000 in the early 1960s (Finney and Houston 1996, 84).

The surf craze following the film *Gidget* was quickly adopted by the mass media and created a wave of surf music and more films. Between 1959 and 1966, Hollywood studios and independents produced up to seventy surf-related films. American Independent Pictures, a production company that concentrated on cheap films, and aimed at the teenage market, produced the most famous of these in a series of exploitation films that included *Beach Party* (1963), *Muscle Beach Party* (1964), *Bikini Beach* (1964), *Beach Blanket Bingo* (1965), *How to Stuff a Wild Bikini* (1965), and *Ghost in the Invisible Bikini* (1966) (Ormrod 2005).

This craze affected not only those in California but also the consciousness and perception of surfing around the world. The sport was beginning to be identified as taking place not in the "wilderness" of the seascape but in the vacation-like and carefree locale of a neighborhood beach town. These films became a new kind of western film, paralleling the social history of the West in which heroic white men and women successfully tamed and civilized the waves of California. Furthermore, the lifestyle portrayed in these films was presented as being attainable and, of course, desirable. As Michel Foucault articulates, new perceptual information leads to new and empirically based theories. Following this framework, surfers began looking beyond what they knew to be their reality, and "saw" new possibilities in a new lifestyle. There was an entire reorganization of the procedure of a new ideology, and eventually a neocolonial tourism industry.

The concept of surf travel, however, wasn't tangibly introduced into the public psyche until 1966 with the creation of the documentary *The Endless Summer*, the first in the trilogy of *The Endless Summer Collection*. Bruce Brown, the director, was initially motivated to create these "documentaries" in response to the commercialization of surfing in the aforementioned Hollywood films, but because of his own presumptuous and entitled view of the role of the West in the sport of surfing, his films became the epitome of a Hollywood fantasy. This series of films widely invoked the idea of colonization in surfers' minds by establishing the "ideal surf lifestyle" as one that involved travel around the globe in search of the "endless summer," an endless holiday. Surfers now wanted to travel beyond American mainland coasts, to discover, name, and experience the frontier of the surf world, and they all wanted to be the "first"—like the two young American surfers portrayed in the film. This cult film became a ninety-two-minute advertisement for a capitalist existence, an ideal adventure of discovery, an experience of the "exotic," a conquest and colonization of "unknown" and "un-ridden" waves in uncivilized places, that could be consumed by brave and determined Western surfers, much like the global explorers of the 1700s. *Endless Summer* films glorified the ideology inherent in the Californian "rhino chasers" of the 1950s for the surfing public.

Figure 1.3 shows the *Endless Summer* film poster from 1966. Joan Ormrod, senior lecturer in lifestyle sports and comics at Manchester Metropolitan University, comments on this poster in her article "Endless Summer (1964): Consuming Waves and Surfing the Frontier":

> The poster copy begins: "On any day of the year it's summer somewhere in the world." It goes on to describe the quest for the perfect wave before concluding with the possibility for audiences to: "Share their [Mike Hynson and Robert August's] experiences as they search the world for that perfect wave that might be forming just over the horizon." The notion of the wave forming "just over the horizon" is a promise of the fulfilment [sic] of the quest but also it holds an echo

FIGURE 1.3. *The Endless Summer* (1966) film poster. This poster was designed around a photograph of Robert August, Mike Hynson, and Bruce Brown by John Van Hammersveld. Van Hammersveld was an art director at *Surfer*. Courtesy of Photofest.

of pushing back a frontier or exploring a new and undiscovered country. . . . It also intersects with the aim of the tourist to discover new
locations in which the "authentic" way of life of other peoples may
be scrutinised. The surfer as constructed through tourist discourse
is a phenomenon which originates from the dissemination of surfing
from the Pacific Islands to America at the beginning of the twentieth
century. . . . In *The Endless Summer* Bruce Brown takes the exotic mys-
tique of surfing the "other" and develops it into a global quest for the
perfect wave. (Ormrod 2005, 41–42)

Michael Shapiro argues that in the twentieth century, the film genre participated in legitimizing Euro-America's "imperial consolidation." He notes
how Western films mythologized and celebrated the "territorial expansion
of Euro-American culture (the westward-moving frontier of violence)"
(Shapiro 2004, 102). Shapiro continues:

Drawing on a remark by the Cherokee artist and writer Jimmie
Durham, who writes, "America's narrative about itself centers upon,
has its operational center in a hidden text concerning its relationship
with American Indians . . . the part involving conquest and genocide,
[which] remains sacred and consequently obscured," [Robert] Bur-
goyne adds that "one of the most durable and effective masks for the
disguised operational center of the nation-state has been the western,
a genre that has furnished the basic repertoire of national mythology."
At a minimum, the classic western films treat, an extended episode of
Euro America's expanding (imperial) sovereignty in politically ambiguous ways. (103)

The Endless Summer films follow the narration style of a western, establishing not the land but the ocean as the expanse of wilderness to discover,
cross, and colonize. Those Others encountered throughout this "spaghetti
western" journey for the perfect wave were identified as "savages," and were
portrayed in the films as caricatures of Africans, Tahitians, and Aboriginals.
Articulated in the *Endless Summer* films is the Anglo-American ideological
project of conquest and dominance, in this case, of select seascapes, by re-
enforcing the West's racial-spatial order in the world (103).

As surfers continued to grow as an identifiable crowd, a capitalist industry followed to meet (and create) their demands, the most imperative being
access to picturesque and uncrowded waves. By the early 1960s, Californian

waves were becoming increasingly crowded, with surf magazines and films transparently boasting about the best locations for waves. Surfer and author Matt Warshaw states:

> Discovery . . . invited ruin. In the late 1950s, when the "perfect wave" designation floated above Malibu like a neon sign, surfers were banging off one another in the line-up like heated molecules. To varying degrees, the same would eventually hold true for the Pipeline, Kirra, Jeffreys, Grajagan, and any other spot renowned and cursed as "perfect." For many surfers . . . the real search for the perfect wave has been less to do with adventure, romance, and the pursuit of new experiences and more with just getting the hell away from what . . . Mickey Dora called "all the surf dopes, ego heroes, rah-rah boys, concessionaires, lifeguards, fags, and finks." Surfers on the road didn't look for anything particularly different. They wanted Malibu (or Kirra, or Grajagan, etc.) without the crowds.[5] (Ormrod 2005, 43)

American surfers began to travel into the Other to reflect their own desires and ideologies, resisting rather than embracing difference as they moved around the globe. What developed was a specific theory of knowledge about the seascape, and thus a specific implication of power in relation to this particular body of knowledge. Mass media and advertisements were set up with the operation of the gaze. Out of this grew a new classification for organizing the surf community, surf-based knowledge, and authority over global surf zones.

This surf safari was celebrated by the Beach Boys in songs such as "Surfin' USA," which lists popular surf spots throughout the world, and "Surfin' Safari," built upon perceptions pushed by films such as *The Endless Summer* and magazines such as *Surfer Magazine*, that promoted travel to distant beaches as both a dream and reality heightened by imagination. *Surfer*, created in 1961, promoted a Californian notion of the fantasy surfing lifestyle. In the words of Fred Wardy, a recognized surfer from the 1970s, "Surfing is a release from exploding tensions of twentieth-century living, escape from the hustling, bustling city world of steel and concrete, a return to nature's reality" (43). This Western definition of surfing emphasizes the concept of escape, of an opposition between here and there, us and them. A specific ideology and consciousness of consuming ideal waves in faraway destinations as a means of "escape" and form of power over both the concrete and oceanic jungles were taking hold in mass-produced surf literature and culture as a capitalist industry.

Shapiro asserts, "For much of the twentieth century, regimes involved in warring violence looked to film as a genre well-positioned to encourage national allegiance" (Shapiro 2004, 117). The *Endless Summer* films helped to establish a national identity for surfers around which an economy could flourish built upon the Euro-American political agenda of cultural governance. Surfers were studied, targeted, and categorized by the growing surf industry. A new tactic of power was systemically imposed on individual surfers. Surfers became individuated by a meticulous measurement of differences between their lives and an ideal surfer's life, imposing homogeneity as well as expectation. It rendered enviable a surfer's designated motivations, spirituality, passions, and physical skills (Alcoff 2005). New modes of expertise emerged and expanded as the knowledges of surfboard shaping, surf photography, and swell forecasting, developed in the surfing industry through mass production. A formulated surf industry was being established, and this industry would travel to Hawai'i and the rest of Oceania as a Western-dominant system capitalizing on an indigenous activity, history, and literacy. All surfers came to be at least mildly subject to these norms of behavior, and the new subject positions of authority produced by this new domain were not imaginable before: the professional surfer, the surf sponsor, the surf photographer, the surf tour-guide, an "authentic" surf lifestyle, ideal wave, remote surf getaway, and the surf icon (Alcoff 2005).

The ideology encouraged and reproduced through the *Endless Summer* films rolled into the Kanaka seascape in sets of waves within the machine of surf tourism. The Hawaiian Islands were flooded in a whitewash of idealism and designated as available for Western consumption as the films identified Hawai'i, already a popular tourist site in the 1960s, as a specific surf Eden. More importantly, this surf Eden was portrayed as a place where the haole are the creators and managers of this paradise, and Kānaka Maoli were not merely peripheral but nonexistent. The result is the perpetuation of an image of Kānaka as existing only for the pleasure of non-Kānaka, as devoid of an independent subjectivity or determination.

The *Endless Summer* films' neglect of native history and the native population in the seascape frees the viewer from any political or social consciousness or responsibility. Instead, the surf public is focused on the determined fact that "no place represents summertime to more people than the Hawaiian Islands." Supporting this conclusion, Brown inundates the viewer with images of haole surfers and beachgoers littering the screen, and white bodies frolicking in inviting waves on air rafts, catamarans, and surfboards. This Hawai'i is envisioned solely through the images of Waikīkī, haole tourists, and haole surfers; it is represented only as a year-round destination for fun and recreation, an eternal land of summertime leisure. Brown narrates, "The water is 75 degrees and so is the air. The temperature only changes about 2 degrees during the year, so unless you have sensitive skin, you can't tell if it's winter or summer in Hawai'i. Hawai'i is truly a land of an endless summer" (Brown 1964). This powerful imagery is profound capital for surfers who do not find value so much in material objects (aside from a surfboard and other surf paraphernalia), but instead in the dream of a perfect wave. Surf culture not only thrives off the acquisition of this fantasy but also finds meaning in the act of imagining this acquisition (Ormrod 2005).

It is worth noting that the *Endless Summer* films emerged from a specific historical and societal moment in American history when a dominant discourse was emerging around leisure, the exotic, and antiestablishment. The sport of surfing came to express the themes of these intersecting discourses reflecting dominant perceptions of American travel, exploration, playfulness, and the myth of the frontier (42). Beginning in the Industrial Revolution the belief was established in the West that happiness could be achieved by the pursuit of well-deserved leisure time, a temporality distinctly set apart from work time. This leisure time was meant for refreshing oneself physically, psychologically, and spiritually, for shedding one's identity and indulging in less responsible and more carefree activities. This notion of leisure time fosters an escapist mind-set of leaving the increasing reality of urbanization for a return to the rural, the wild, to a lost paradise (42).

This perception was carried through to the 1960s and '70s surf community and created a consciousness that authorized a process by which a specific identity of the surfer was formed and metamorphosed through the specific terrain of the West. Societal relations formed between the surf tourist and the native that structured a power grid based on the distinction between "us" in civil society and "them" in apolitical and peripheral (leisure) space. In turn, this grid produced the knowledge of expectation and entitlement portrayed in the *Endless Summer* films, in which the native existed outside of society. There was a distribution of the economy of leisure through which a new discourse of knowledge about surfing was produced—a new "truth" of the sport that entitled the surfer the right to any wave he or she could "discover"—that gave impetus to the development of surf tourism. It established a political relationship between the surfer and the Other.

Predominant themes of the *Endless Summer* films include the known/ unknown, named/un-named, exotic/familiar, and crowds/isolation (Ormrod 2005). All binaries in the surf tourism industry are interrelated; the known is named, familiar, and crowded, while the unknown is unnamed, exotic, and isolated. The arrival of the American surfer to the unknown, exotic, and remote surf destinations is a continuation of the colonization and "civilizing" of the wilderness, the unshaped, the unnamed, and unknown surf spots in Hawai'i. Intentionally omitted are the indigenous histories swimming inside ke kai.

(Re)Moving Places

All sports operate in spatial metaphors such as fields, courts, tracks, the sky, and other boundaries or nonboundaries. Surfing, as a sport, operates within the spatial metaphor of the ocean, which from the early 1900s through the 1960s was perceived by Americans to be one of the final frontiers. To Americans, the ocean had no visible or stable boundaries, as in a football field, no powder lines or erect metal posts to determine success of scoring. Surfing was more of an expedition in the wild: men equipped only with a board, or "elephant gun," as big-wave boards were called, and a pair of swim trunks. American surfers were among the "cowboys" mastering and civilizing territory. These early "rhino chasers" imagined themselves as big-game hunters, "roaming around on the rugged North Shore [of O'ahu], trying to bag the biggest and wildest waves of the day" (Coleman 2001, 46). The contemporary surf tourism industry has capitalized on this idea of an exotic ocean space still out there, waiting to be discovered. A surfer no longer needs to limit himself or herself to local or even national horizons; there are numerous "new" ocean spaces in which to surf. The sport's boundaries have expanded to include Antarctica, the Galapagos, Nova Scotia, Madagascar, the Amazon River, Lake Superior, and even traveling wave pools on cruise ships.

As the activity of he'e nalu travels, the spatial metaphors of surf "spots" also move: marching waves, circulating currents, crawling sand. Even images of surf spots move through photographs, films, stories, songs, and art. Bodies within these changing locales have also moved, touring new spatial possibilities within the ocean: the movements of spatial metaphors, images, the physical sea, and human bodies are all potentially political movements capable of either solidifying or displacing cultures and identities. In Hawaiʻi, for instance, moving Kanaka bodies remember and re-create connections to ancestors, while touring surfers bring their own ideologies and modes of dwelling to Hawaiian waters, creating their own stories, knowledges, and theories about the spaces they encounter. Spaces themselves are created and destroyed as the enactment of heʻe nalu moves.

Surfers create spaces by moving in curves and within improvised syncopations that are both smooth and at times violent. An architecture of space is built that flexes with every extreme movement made by the body, surfboard, and ocean, altering the dimensions of the playing "field." In Hawaiʻi, the traditional *kapu* (taboo, prohibition) system limited space for Kānaka Maoli, as did the geographic division of the ahupuaʻa system, so ke kai was not an unclassifiable, amorphous place. The influx of the surf tourism industry's narrative into Hawaiʻi's oceanic space, however, confined the creative realm of possibility for Kānaka Maoli in a different way. Movement was limited by foreign bodies crowding surf breaks, channels, and beaches, leaving no "space" for autonomous Kanaka movement. This phenomenon is nicely articulated through David Winner's description of Dutch soccer:

> Total Football was, among other things, a conceptual revolution based on the idea that the size of any football field was flexible and could be altered by a team playing on it. In possession, Ajax [the Dutch team] — and later the Dutch national team — aimed to make the pitch as large as possible, spreading play to the wings and seeing every run and movement as a way to increase and exploit the available space. When they lost the ball, the same thinking and techniques were used to destroy the space of their opponents. They pressed deep into the other side's half, hunting for the ball, defended a line ten yards inside their own half, and used the offside trap aggressively to squeeze space further. (Winner 2000, 44)

The Dutch created space for themselves in which to maneuver, and shrank space for their opponents. The surf tourism industry exceeds this aggressive and brilliant tactic on the playing field of Hawaiian waters, expanding their surf territory by attempting to not just shrink but erase indigenous *places* (spaces with great historical, cultural, economic and social significance). The surf industry has reduced genealogical and cultural space in the Kanaka sea with an imperial understanding of history. The result has been an overshadowing of Kanaka claims to land and sea, and the expansion of American maneuverability.

For example, surf media has created the myth that American surfers helped to recover the "lost" sport of surfing in Hawaiʻi, taming the native waves on the islands. Californian surfer and filmmaker Bruce Brown asserts in *The Endless Summer* that 1958 was the "first time" the big waves of Waimea Bay on Oʻahu were ridden.[6] Stewart Holmes Coleman continues the imperialistic narrative, writing in *Eddie Would Go* that in the 1950s, "during those first expeditions to the North Shore, men like John Kelly, George Downing, Greg Noll, Pat Curren, Peter Cole and Fred Van Dyke would drive across the island in their old jalopies, camp on the North Shore and surf pristine and unridden breaks that had yet to be named. Of course, they suspected that the ancient Hawaiians had probably ridden these waves, but it seemed impossible with the long, heavy planks of wood they used to ride" (Coleman 2001, 45).

Kānaka Maoli undeniably rode and named the waves at Waimea Bay and elsewhere. A *kau* (sacred chant) of Hiʻiaka, for instance, tells that Piliʻaʻama, the *konohiki* (the headman of a land division under the aliʻi) of Ihukoko (the landing up the Waimea River, when the rivermouth was still open), enjoyed surfing both the point and the shore break at Waimea. Elspeth P. Sterling's and Catherine C. Summers's *Sites of Oahu* states that this kau is from the moʻolelo, "Hiiaka-i-ka-poli-o-Pele."[7] The section of the kau referring to surfing at Waimea reads,

> The sea sprays up over the sand,
> The yellowing sea of Pupukea,
> Yellows the leaves of the ilima [native shrubs, bearing flowers]
> With its sprays.
> This is the way to Kapi,
> This, the trail to Pilia'ama.

> At Pupukea they spied Piliaama fishing and as they watched Hiiaka
> called:

> O Piliaama fisherman of the cliffs
> Who surfs to the mouth of the stream of Ihu-koko
> Who catches aku fish at Kaipahu,
> What fish are you catching now?
> (Sterling and Summers 1978, 144)

Kanaka historian George Pooloa wrote a piece titled "Na Pana Kaulana o ka Aina o Oahu: Noted Places on the Island of Oahu" that also confirms that Kānaka surfed the waves at Waimea.[8] *Sites of Oahu* has a translated excerpt from this piece under a section titled "Surfing at Waimea":

> She left her friends and turned homeward to Kahuku. At Waialua she encountered a wind and rain storm. The water poured down the hillsides and washed out the trails but she cared not at all.
>
> At Waimea (there were no bridges in those days) she went along the shore in the storm. The river of Waimea almost overflowed its banks and the beach was full of people. In the olden days the natives of Waimea enjoyed riding the huge surf and up into the stream when it was slightly swollen. This sport was called the wai-puʻeone.
>
> Thus our traveler crossed Waimea on a surf board without encountering trouble. This was said to be a favorite sport of the olden days and the boards did not sink when managed with skill. (129–130)

Place names have also been effaced with the proliferation of surf tourism in Hawaiʻi. Sunset Beach on Oʻahu's north shore, one of the most famous surf spots in the world, was named Paumalū by Kānaka Maoli. Paumalū is also the name of the ahupuaʻa in which the break sits, a common practice of naming in Hawaiʻi because reefs were and are perceived to be extensions of the land; thus the area's names correlated. *Sites of Oahu* quotes Henry Kaina in Gilbert J. McAllister's *Archeology of Oahu*, explaining the significance of the name Paumalū, a site choice for catching squid:

> At one time there lived on the island of Oahu a woman who was noted for ability to catch squid, of which the chiefs of high rank were fond. If there was anyone who could catch a lot of squid that person was in great demand.
>
> One day a great luau was to be given by a chief, and he wanted some squid. He sent some of his men in search of someone who could catch squid. They brought the woman to him. He told her he wanted squid from a certain reef and asked her if she could catch some for him. She said she could catch all he wanted.
>
> She went down to the beach at the place designated by the chief, but before she entered the water an old man met her. He told her the rules of the place: she was supposed to catch only a certain number and when she had caught them to go home, or something would be

sure to happen to her. She called for her daughter who had followed and told her to come with her into the water. Another thing the old man had said was for her to go home when she said she would and not to stop for anything. The lady caught all she had been allowed by the old man, but she kept on fishing until she had more than she could handle. She sent her daughter to the shore with half of the load and told her she was going home, but instead she remained, for she saw a huge squid she wanted to get. Just then a large shark came and bit off her legs. She yelled for help. Her daughter came to her rescue, but too late. She died from the loss of blood and the shock.

When the people examined her later they found one deep gash on her right arm made by one of the shark's teeth. They then knew that it was done by a shark who guarded that particular reef. After that incident they named the place Paumalu, which means, "taken by surprise." (145)

Another moʻolelo about Paumalū tells how this was the stretch of ocean where the prince of Kauaʻi, Kahikilani, came to prove himself in the 1700s as a surfer:

Long ago there lived on Kauai a chief who was very fond of surfing. He had won every surfing contest on his own home island and now came to Oahu to try his skill. As the surf at Waikiki was not to his liking, he went on to the Koolau side of the island. There he found just what he wanted.

While he was surfing he noticed some birds circling about him. One old bird in particular would fly a short distance away and then return to circle about him as if urging him to follow. He did so, and the bird led him into a cave where he met a beautiful girl who had fallen in love with him as she watched him surfing and had sent her pets, the sea-birds, to lead him to her. She asked him to become her husband and he accepted her proposal. Each morning before he left her for his favorite sport she made him two lehua [flower of the ʻōhiʻa tree] wreaths to wear, one for his head and one for his neck.

For a long time they lived thus happily until one day as he came ashore from surfing, another girl greeted him and threw about his neck several strands of the golden ilima. The old seabird flew home and reported to his mistress what he had seen. When she saw her lover returning with the ilima wreaths about his neck in addition to the lehua strands which she had braided for him, she was very angry and called upon her ancestral gods [*aumakua*] to punish him. As he ascended the hill he felt his body becoming heavy and, as he turned to look once more at his beloved surfing beach, there he remained transfixed in stone and is so to this day. (Sterling and Summers 1978, 146)

There is another version of this moʻolelo in *Sites of Oahu* as told by Rayna Raphaelson in *The Kamehameha Highway: 80 Miles of Romance*. This version's narration includes much more oceanic imagery, and details the waves at Paumalū: "Now there are waves, but the waves on this beach come in from the sea in a way that puzzles all riders of surf. They do not come sweeping in straight, long swells. Some distance out, there is a lull and a change. Not many men can ride these waves. And from long ago, they tempted the stranger who came from Kauai" (146).

Today this area is predominantly called Sunset Beach. Losing the name Paumalū washes away the memories of prince Kahikilani, Hawaiian lessons of pono fishing in the area, and a knowledge of how the waves break along this shoreline. Sunset Beach becomes an oppressive reference in Hawaiʻi.

The surf spot Velzyland was also renamed, after a haole surfer and surf film producer, Dale Velzy. On one of Velzy's Hawaiian explorations, he and director Bruce Brown "located some surf just past Sunset Point [Paumalū], in the land division of Kaunala. Recognizing the advertising potential of this unnamed break, unnamed to them, they called it 'Velzyland' after their sponsor" (Clark 1977, 130). This colonial act discarded the Hawaiian spatial history along this coast (the term "spatial history" defers to Paul Carter's definition, which speaks to history not as a chronological narrative but as a poetics of movement through space), a relationship created by the indigenous people who had lived on, surfed, fished, and cultivated these reefs for centuries. Clark informs us that the surf spot known as Velzyland was called Kaunala Beach by Kānaka ʻŌiwi; it means "the weaving," although the significance of this Hawaiian name is unknown today.

Place names are formed from an accumulation of information gathered and insight shared from generation to generation. Clark elaborates,

If you think about the earth, the earth came without names. Originally, it was just the planet earth. So place names come from people, and people name places because those places have some importance to them. . . . If you look at Hawaiian place names, if you know the moʻolelo, the story behind the name, and why it was named what it was named, then you can see what the value was and why it was important to the Native Hawaiians. . . . That speaks to the culture, it speaks to how the beaches or shorelines are used, and it tells you exactly what people were doing there and why. (Clark 2007)

There is a form of symbolic appropriation through the renaming of surf spots in Hawai'i, claiming authority over waves, as well as land and history. Carter believes that the establishment of place names was a traveling tool for European explorers, part of the process of conquest rather than a result of discovery. This act of (re)naming places attempted to incorporate this "new" terrain into the European imagination, defacing indigenous spatial history in the process. "In fact," Carter says, "the place names navigators and explorers bestowed were frequently periautographical. They preserved in a cryptic fashion a record of incidents during the voyage, as well as evoking the network of scientific, professional, and political allegiances that had either informed or sponsored the passage or whose patronage protected its interests" (Carter 2009, 50).

Renaming muffles and silences those indigenous political structures, cultural relations, and the imaginative, technological, and architectural content associated with places, all of which are carried in their names. Names hold the human freedom of a people, their ability to construct identities from their cultural libraries about the spaces they maintain and from which they are sustained. The place name Kāhala on O'ahu speaks to the accumulated fishing knowledge gathered by Kānaka Maoli from this region over the years. Clark recounts, "The full name kāhalai'a means 'the amberjack fish.' Another possible definition of the name that is associated with the ocean comes from Kanaka historian David Malo, who wrote that kāhala was a method of catching sharks with a hook and then using a net of very strong cord to ensnare them. Advocates of this interpretation point out that Lae o Kūpikipiki'ō (Black Point), which adjoins the Kāhala area, has long been known among local fishermen as a shark ground" (Clark 1977, 38).

The sentiment of ownership by Western ocean "explorers" who consistently claim and rename their "discoveries" continues to occur in the world of surf tourism. *Water* magazine reflects this sentiment in the opening of its 2005 issue with a photo taken from the window of a twin-engine propeller plane of a small, isolated atoll with waves breaking along one coast and a lagoon surrounding the other. Sunny Miller, a writer for the publication, captioned the photo: "This is your captain speaking . . . to the left side of the aircraft is a South Pacific dream. ENJOY" (Miller 2005, 15).

Dane Larson, surf journalist, describes the evolutionary phenomena:

> While the Conquistadors may have been searching for their version of wealth and power, the traveler today searches for riches in the form of unspoiled beauty. A fundamental law in our modern travel-happy world: desirable destinations will be discovered, and then colonized by paradise-seeking imperialists of the economically advantaged order. Nowhere is this more apparent than with surf destinations. . . .
>
> Surfers themselves are modern day explorers. The singular motivation? To discover high-quality, deserted waves. Perhaps no other group with the possible exception of ocean sailors is as motivated to explore remote, uncharted coastline and waters. The reward: to realize hidden, undiscovered treasure in the form of waves. Your own private surf hideaway, reserved for you and your closest friends. . . .
>
> Enterprising locals and/or foreigners then begin to set up small businesses, such as restaurants and lodging. These amenities, in turn, open the area up to a whole new set of vacationing surfers, and the entire process accelerates rapidly. Next come the surf camps, surf tours, advertising, web sites, additional media exposure, and . . . you have the potential for the full-blown surf ghetto. (Larson 2005, "Is Surfing Etiquette Dead?")

Some of the world's best surfing waves have actually been purchased. Oceanic regions can now be privately owned as tourist attractions, as seen in Fiji on the island of Tavarua, and on surf charters that "own" exclusive rights to specific waves in the Menatawai Islands in Indonesia. Australian surfer Martin Daly provides the archetype for this new surf identity. Daly is credited with pioneering one of the most popular and elite surf adventures in the Mentawais, and, more significantly, with pioneering a new way to travel, to identify oneself as a surfer, and to surf. For almost a decade, Daly and a handful of his employees had managed to keep his "discovery" of perfect, uncrowded waves off the Menatawai Islands a secret from the rest of the international surfing community. In 1992, however, photos were leaked to the press from one of his private boat charters, setting off a whirl of desire and opportunity in the minds of both surfers and surf entrepreneurs. Today the Menatwais are bustling with over forty boats operating offshore surf charters, and there are a growing number of land-based surf camps (Parkinson 2008).

Daly, who exudes a sense of ownership over the region, isn't pleased with this explosion: "It's kind of like watching a bunch of seagulls fighting over a chip" (59). In an egocentric gesture, Daly has spent the past five years voyaging around the world in search of a "new" perfect wave that he can once again claim as "his" discovery. The group of islands that house Daly's new surf utopia are located in the Western Pacific, in the Northern part of the ocean, once again subjecting the Pacific seascape and indigenous peoples to a neocolonial endeavor, a game of discovery and power. Daly boasts that this region is "very hard to access," mentioning only once in his extensive interview with *Water* magazine, that there are indeed people living on these islands, "some" of whom surf. The discovery of this new wave is akin to that of Columbus, including the colonial intentions and inevitable inequalities between colonizer and native in such an egotistical encounter. It is a culture-centric approach to existing and perceiving the world. Daly boasts that he'll be spending much time in this "new" region of the seascape: "*Trader 4* [his charter boat] has been there for a few years, and *Trader 1* has been up there twice" (65). It turns out this "new" wave is in the Marshall Islands, where Daly has subsequently taken out a ninety-year lease on one of the forty-five-acre islands where he plans to build a resort on land.

This neocolonial ideology is illuminated in how Daly posits his "dilemma" in surf travel and discovery, which is actually the universal dilemma for all surfers: he wants to bring surf tourists to these idyllic places for capitalistic purposes, but he also wants to keep the lineups uncrowded for his own pleasure. "When it's quiet out there, we're stoked because we get more waves, but that means less income. Now that I think of it, it's a win-win situation, really. You're either making money or surfing perfect waves alone" (67). The dilemma for Daly is whether or not he's "revealing" these "secret" spots to the surf community. It's *how* these places are exploited, whether by him alone or along with forty other "pioneers." The ideology remains one of discovery, dominance, and fantasy, which activates a neocolonial institution.

Daly goes so far as to divorce himself from the rest of the surf capitalists, stating, "There are people out there [in the Mentawais] who own and operate boats and live the surfing lifestyle. And then you have this non-surfing, capitalist pig element that is just there to take money out of our thing, and they're making things really hard right now. They're trying to impose taxes on every surfer who goes there, they're colluding with the Indonesian and Mentawais governments, trying to put up fences and get exclusive rights for all the waves. Surf camps are trying to get rid of charter boats" (69). His unconvincing argument is that he has invested his dreams and life in the "discovery" of waves, and he does not appreciate anyone creating a wake in his territory. He fails to mention any consideration about the lives of the indigenous people and their investment in their own resources. Daly continues, "I have to go to a meeting with the boss of the Mentawai Islands and others on Sunday to try to straighten things out. They had a demonstration against me in Padany last week, calling for my deportation because I disagreed that a boat should have to work underneath a land camp and give them a large chunk of their profit for doing nothing" (71). What Daly conveniently omits is that he is a visitor in this land, that the local government and people have the right to impose taxes that would help the local communities, not just the other land-based surf camps. Daly's sense of ownership and entitlement because he "discovered" this secret and new wave is revealing of a larger political and social dichotomy between the West and the Other.

This sentiment of territorial conquest is also found in Western language about the act of riding water. Surfing is discussed through the terminology of "ripping," "shredding," and "killing" waves. *Surfer Magazine*'s April 2006 edition described surfer Pancho Sullivan and his skills in these terms: "At 6-foot and nearly 200 pounds, Pancho attacks big waves like a varsity high-school wrestler. . . . Then he proceeds to vaporize the lip [of the wave] with a single broadsword sweep that leaves steam hanging in the air" (102).

Spatial histories have not been completely washed over in Hawai'i, but the recoding of Hawaiian ocean spaces is part of the displacement, part of pushing out and pushing aside the indigenous. The social and political order once provided by the performance of he'e nalu no longer gets played out within the traditional systems of kapu. The struggle for order, pride, power, and *mana* (spiritual power) is now largely enacted within a foreign system established by the contemporary surf tourism industry. It has expanded into a larger struggle for neocolonial order, a struggle that is taking place in a new territory, running off the land into the territory of the ocean. Ocean "territory" and access to the resources involved in surfing are the new battlegrounds of surf colonization. Joan Ormrod reflects on this phenomenon,

> It [surfing] is a culture based upon consumerism. A quest for the perfect wave could not have been made by any other society or subcultural surfing group than American, specifically Californian, at this time. All the conditions for such a journey were in place within this

specific time and cultural moment. In addition to being a journey of exploration and adventure, it was a journey of colonisation and incorporation of other surfers and waves into American surf culture. It affirmed America's dominance of global surfing at the time. (Ormrod 2005, 49–50)

This is particularly true in Hawai'i, America's fiftieth state, where private and corporate development, disguised as progress, naturalizes the marginalization of Kānaka Maoli both in and out of the water. Surf tourists are not expecting to see, nor are they shown, the economic and political inequalities inherent in Hawai'i's niche market because the dominant movement of Western surfing is a form of travel, or "touring," that creates specific orientations and ethical trajectories of constructing and consuming others (Soguk 2003). Many jobs available to Kānaka in the surf tourism industry require "performance" as a beachboy, surf guide, contest surfer, or hotel or restaurant worker. More than a demand for performance, surf tourism commands an absence of the native unless he or she is operating within the system.

Surf tourism becomes an economy, as a flow of capital in and through Hawai'i, Hawaiian subjects, and tourist consumers organize these subjects as economic units (Halualani 2002, 141). Kanaka scholar Rona Tamiko Halualani gathered that roughly 60 percent of Kānaka Maoli held tourist-sector-related jobs (hotel, service, restaurant, transportation) in 2002, which accounted for 25 percent of the entire workforce in Hawai'i. Yet these jobs paid less than the state average at the time ($20,000). "As money seems to pour out of and into Hawai'i, local Hawaiians materially struggle in an industry marketed on their history, their images, and their colonialist depression," Halualani writes (144). The tourism industry is a multibillion-dollar business in Hawai'i. All of the profit brought in by the surf tourism industry "confuses" the "historical-spatial boundaries between culture and commerce." Kanaka seascape, a place of profound ontological and epistemological significance, is transformed into a mass-produced commodity available for consumption and to be experienced. Surf tourism perpetuates what Halualani calls a "looseness of confusion between culture and commerce" (175).

Contemporary surfers, like most tourists, have slipped onto the path paved by colonialism, which has exempted them from witnessing or having to concern themselves with the geopolitics and marginalization of indigenous populations just ashore of annual multibillion–dollar-grossing surf camps and boat charters such as Daly's (Barilotti 2002). Surf tourism has become an industry that takes from a culture what serves it: waves, idyllic weather, access, "remote" locales, and adventure, without much insight into the impact it has, or level of reciprocation it offers to native populations living in their surf utopias.

Isaiah Berlin, social and political theorist and philosopher, tells us, "Utopias have their value—nothing so wonderfully expands the imaginative horizons of human potentialities—but as guides to conduct they can be literally fatal" (Berlin 1998, 12). He believes that the notion of the perfect whole, in which all good things coexist, is not merely unattainable but conceptually incoherent. It is incoherent because supreme values are relative to time and culture. Berlin cites the example of Niccoló Machiavelli, who asserted that Christian values—humility, acceptance of suffering, unworldliness, the hope of salvation in an afterlife—are bound to be trampled on by the ruthless pursuit of power in the attempt to re-create and dominate the republic (12). These two moralities are not placed hierarchically, but are simply incompatible, although Machiavelli knows which he prefers. Berlin writes,

> So I conclude that the very notion of a final solution is not only impractical [as time produces new problems to solutions] but, if I am right, and some values cannot but clash, incoherent also. The possibility of a final solution—even if we forget the terrible sense that these words acquired in Hitler's day—turns out to be an illusion; and a very dangerous one. For if one believes that such a solution is possible, then surely no cost would be too high to obtain it; to make mankind just and happy and creative and harmonious for ever—what could be too high a price to pay for that? (12–13)

The imagined American utopia in which one can discover the perfect wave and an endless summer is not necessarily a final "solution," but it is a final utopia and a desire that has profound effects in Moana despite the seemingly benign intention of riding the waves. Perhaps the insertion of the word "perfect" is the culprit. This advertising slogan and the commercialization of the ocean is what coaxes surfers into purchasing the perfect Pacific wave, the oceanic surf fantasy.

However, as Teaiwa notes, tourism statistics can tell only so much about tourism in Oceania, and can tell even less about relations between the tourist and the Native, a nonabsolute relation (Teaiwa 2001). Similarly, while the image of and certain spaces within the Hawaiian ocean have been hijacked by the colonial project of surf tourism in Hawai'i, the *enactment* of he'e nalu has not. Since annexation, he'e nalu has been profoundly disciplined by foreign political and economic interests, but it has not been "colonized." Claiming he'e nalu as colonized implies that the activity is no longer socially, politically, or spiritually significant to, nor is it controlled by, Kānaka Maoli. On the contrary, he'e nalu has always found pockets of resistance to increasing commercialization and commoditization. He'e nalu remains a contemporary way of connecting to an indigenous ontology and epistemology about the interconnection of passages leading out from an origin within.

Breathing Underwater

A Kanaka surfer can "breathe" underwater. When a wave pushes her lungs down against a rocky bottom, and she is caught in a liquid vortex of churning circles, she has learned that she must remember her human "dive reflex," an automatic reduction in heart rate and oxygen consumption stimulated the instant her face hits the water. As humans, our bodies are made to live on land, but we were born from a watery womb as aquatic beings extracting oxygen from fluid: human brains are 75 percent liquid, and our bones are 20 percent liquid; we are hairless land mammals like dolphins and hippopotamuses; and our bodies are insulated with subcutaneous fat, enabling buoyancy; and human blood is more similar to salt water than it is to fresh (Farber 1994). Our bodies are made to continually interact with water. The submerged Kanaka surfer must relearn to breathe under the sea, becoming a pelagic creature once again, as the primitive webs between her fingers and thumbs paddle her back toward the glowing light above the surface.

Time slows as a surfer is submerged under the power of the wave, hair and lashes whipping in a familiar movement. Seconds feel like an entire night's dream as she swims through the watery darkness. Her body is trained to resist the reflexive messages conjured in her mind that she is out of air, causing panic and quickening the consumption of her limited oxygen supply. She knows she has three or four seconds of breath stored in her lungs from the deep inhalation taken from her diaphragm before going under. This is a psychological test as much as a physical one as she swims upward.

Exhaling the toxic carbons of colonialism and finding an internal supply of self-determination stored in her own historical identity allows the

FIGURE 1.4. Surfer paddling near lava flow, from the *Honolulu Star Bulletin*, July 10, 2008. Photo by Kirk Lee Aeder. Courtesy of kirkaederphoto.com.

Kanaka surfer to cleanse, re-create, and reconnect from under the crush and spin of the surf tourism industry. Being held under is part of the engagement in heʻe nalu, and the drowning attempts made by capitalism and colonization are part of a contemporary Kanaka reality. To navigate around the chaotic churn of colonialism, Kānaka Maoli have always and continue to draw upon the culturally aquatic marrow in their bones for guidance. There are distinctly Hawaiian ways of seeing space and place, even from under water. Professional Kanaka surfer C. J. Kanuha, for example, made headlines across the world in April 2008, and was interviewed by ABC's *Good Morning America*, British Broadband, and other media corporations, for paddling his stand-up surfboard within twenty feet of lava flowing into the ocean from Kilauea's eruption at Waikupanaha in Puna (see figure 1.4).

What appealed to the media and surf industry was the danger and novelty of the act. Kanuha, however, underwent respectful training for this event as a way to visit with Pele, the Kanaka volcano goddess. His movement was primarily cultural. When Kanuha reached an area of the lava that had formed a black sand beach, he went ashore and prayed to his goddess, offering her *ho'okupu*, a gift of respect. Kanuha was employing his oceanic knowledge to strengthen his indigenous ontology and epistemology independent of the cultural location of the corporate sponsors supplying his board, paddle, and, most likely, a pat on the back for the attention. What took place was an interplay between the surf industry and a Kanaka space within it.

Kanuha is a major participant in the industry. The website for his surf school, Hawaiian Surf Schools and Adventures, lists his credentials: he was featured in the surf videos, "Snap One," "Snap Two," "Blueprint," "Common Thread," and "Fifth Symphony Document"; he has been featured in *Surfer*, *Surfing, Waves, Transworld Surf, Outside, Men's Journal, Sports Illustrated, Free Surf, Surfing Europe, Hana Hou, Surfer's Path*, and *Surfer's Journal*, among others; and his sponsors include Etnies Shoes, Nixon Watches, JEN Optics, On A Mission, Sector 9 skateboards, Surf Prescriptions surfboards, and the Kona Brewing Company. Sitting so deeply inside the surf tourism industry, Kanuha and many other Kanaka surfers find themselves successfully balancing two distinct epistemologies. Kanaka uses the modern surf industry as a vehicle to experience and (re)affirm an indigenous identity connected to and emerging through ke kai.

Contemporary Kanaka spaces are both indigenous and modern, both spiritual and capitalistic. Tourism in Hawai'i is an intricate ensemble of imposing relationships between Kānaka Maoli and the tourist, sometimes requiring Kānaka to become Other and at other times demanding that they become more native—not colonizer, traveler, and so on (Teaiwa 2001). The relationship between Kānaka 'Ōiwi and surf tourism is an interactive labyrinth; there is always an exchange of enablement and disablement between agents and events. The experience of surf tourism in Hawai'i cannot be universalized as neocolonial, constructed as an entirely negative or inescapable system. Colonial power can be exaggerated, diminishing the extent to which colonial histories were shaped by indigenous resistance and accommodation, as was the case with the original Waikīkī Beachboys.[9] Colonial ideologies were and are more variable and ambivalent than most dominant anticolonial discourses reflect. The "neocolonial" surf tourism industry remains neocolonial, but this recognition does not assume an unlimited power.

There is, however, purpose in recognizing the neocolonial aspects of the industry. Recognition enables the deconstruction of dominant narratives and structures that prevail, illuminating how Kānaka Maoli can and do sit inside, outside, and between them. The economic, political, and theoretical implications of this interactiveness are significant. Kanaka culture and identity are not merely objects of surf tourism; they are also the subjects (Teaiwa 2001). Surf tourism becomes a fertile sea in which Kānaka Maoli can negotiate and define modernity as their identities philosophically maneuver over and around designated roles, as Kānaka swim into autonomous and resistant waters. Furthermore, just as modern Hawaiian watermen and -women are both inside and outside, empowered and, at times, disempowered within the system, the neocolonial process of surf tourism is also nonlinear in action or reaction, requiring that the ideological structure of the industry be addressed rather than unilaterally condemned.

Surfers are organizing into visible political and environmental groups, altering how the surf community approaches the ocean. A new ideological outlook that veers away from language about discovery, conquest, and control could empower indigenous spaces in the seascape while allowing for the continued functioning of a surf tourism industry. In looking toward solutions, Steve Barilotti asks the critical question, "So how does one create sustainable surf tourism that benefits the surfer, the local, and does as little damage to the environment as possible?" (Barilotti 2001, 96). Answers to this question are being discussed in both local and global surf communities, and between them. A number of organizations have formed in an effort to keep the ocean environmentally, socially, politically, and culturally healthy: the Groundswell Society, an ad hoc surfing think-tank that audits and analyzes the culture and ethics of surfing and surf culture; the Surfrider Foundation, a nonprofit environmental group (United States); Surfers Against Sewage, a nonprofit environmental campaigning for clean, safe, and accessible recreational waters (Great Britain); SurfAid International, an international health care service working in surf-rich regions such as Indonesia (Australia); and Save Our Surf (SOS) represents a Hawaiʻi-based grassroots group, formed in the early 1960s, mainly by young surfers, in response to increased environmental degradation and overdevelopment in the islands (particularly the destruction of a favorite surf spot in Ala Moana) by dredging to expand the beaches of Waikīkī for tourists.

The shift away from multinational interests toward local agency is particularly challenging in Hawaiʻi, however, because, since statehood, American tourists have claimed both a legal and a psychological entitlement to the Kanaka kai. Local politicians and corporations and many residents perceive the surf tourism industry as a positive and noninvasive industry because it is publicly linked to economic prosperity. While there is truth to this narrative, there are also hidden costs to Hawaiʻi's residents: ecological destruction, low wages for the majority of industry workers, water consumption, traffic, displacement from land/ocean, and cultural commodification. Increasing local agency within the surf industry would likely increase responsible ecological and culturally respectful corporate conduct, and it would help to keep the bulk of economic profit from the industry inside the islands. Supporting local surf businesses in Hawaiʻi, such as Kealopiko and Country Feeling Surfboards, and environmentally and culturally minded surfers, such as Jack Johnson and Duane DeSoto, works toward this end.

The ambition is to move surf instructors, for instance, from operating inside the ideology of the surf tourism narrative, and offering commercial lessons to a mostly tourist market without first instilling an awareness of the critical relationship between the surfer-to-be and the ocean, toward an alternative relationship with ke kai. Lessons today consist of technical how-tos: lay here, paddle this way, kneel and jump to a standing position like this. Brief backgrounds on swells, currents, reefs, and winds might accompany an introduction, but for the body to be truly affected by the sea, it must listen, observe, and sit inside the waves. The ocean should not be approached as merely a recreational and consumable space. Heʻe nalu needs to become an activity endowed with a purpose beyond standing on a surfboard, even for a tourist who will never have the opportunity to visit the sea again. Heʻe nalu should come to involve an engagement between the mana of the ocean and an individual's own epistemology; how she connects her ways of knowing the world to her body's interaction with the waves she is engaging on that one day in Hawaiʻi. This connection must become available to every surfer regardless of age, gender, religion, nationality, or ethnicity, and regardless of how long this connection remains part of his or her consciousness. Even if fleeting, the potential of heʻe nalu as an oceanic literacy puts the body back into the places it travels through, even while taking place inside a capital time. Time and space can be rhythmically reconceptualized—not deleted or overturned but rediscovered inside colonial structures (Carter 2009). A discussion should ensue about how to allow for a reconciliatory coexistence that honors the indigenous knowledge of heʻe nalu within the commercial industry. At stake is the global cause of a "just society" within the ethical notion of a surfing "community."

For contemporary Kānaka Maoli, heʻe nalu is a process of acculturation, having adapted to Western ways and materials such as fiberglass boards, plastic leashes, traveling on airplanes to distant waves, wearing nylon printed suits with corporate brands, and using Internet access to determine wave conditions. Heʻe nalu continues to attune the body to the ocean's liquid pulses, which create earth passages allowing for an interplay of ideas and perspectives. Identity rolls in and out of the world's shores, tasting and feeling diverse lands in a "dance of the intellect," as Carter calls it. Seascape epistemology paints a world in which identity does not exist but is always emerging and moving inside the places it inhabits. The beauty and power of seascape epistemology comes from the impermanence of the messages written in the sea. The transitory nature of oceanic literacy ensures re-creation within the initial memory of the knowledge experienced in the waves and on the sand (Carter 2009).

As a conceptual way of knowing and being, seascape epistemology offers Kānaka Maoli a dynamic link to place, resisting an essential connection to land and sea in a modern era of Kanaka diaspora and travel. Oceanic literacy becomes a phenomenological tool for Kānaka, a political literacy that can reimagine possibilities through sensations felt not merely by the surfer as an isolated figure but all of the possibilities included in her image: the waves, the sand, the reef, the fish, ancestors, and future generations. These sensations radiate out, reaching those watching on the shore, affecting and pulling them (even tourists) into the experience. Heʻe nalu is an enactment by which energy is released, digested, and then rereleased in an incessant cycle of rejuvenation and (re)connection.

A surfer's car smells of decaying ocean water. All of the microorganisms brought home on her towel and dripping from her fiberglass board suffocate in the confines of this metal box, with its windows rolled up, creating an atmosphere devoid of nutrients. The water doesn't "die"; it transforms, partly into rocks of salt, partly evaporating into the air as vapor that will soon rain down, back into the sea. Feet transport sand from Mākaha to Kaimukī, lodged in the seams of leather seats and between wool loops in the living room rug. The seascape moves beyond its own body to sit in our human places just as we extend our bodies to sit in it. Kānaka Maoli and the ocean are always connected, even outside the time and space of ke kai, through the movements of surfers, as well as of fishers, voyagers, paddlers and divers, who physically and emotionally take part of the seascape with them as they travel. These movements enable the ocean to articulate its message of union between human beings and the jellyfish, the crests and caps of swells, and the star's dust. This is the potential that Kānaka Maoli can tap into through the oceanic literacy of he'e nalu, a corporeal literacy rather than a cerebral one.

OCEANIC LITERACY A Politics and an Ethics

Oceanic Literacy

Sitting offshore, in the Hawaiian sea, near the Mōkapu Peninsula, which lies within the ahupuaʻa of Heʻeia in the district of Koʻolaupoko on Oʻahu, a Kanaka *kino* (body) becomes particularly expressive through kinesthetic and emotional awarenesses and experiences of knowing and being that are fluvial, that flow and sway with the rhythms of waves. Bobbing in long fingers of swells, this Kanaka surfer sits beside an ancestor, Kūʻāu. They both wait for a breaker in the sea at Mōkapu. Kūʻāu is a *pōhaku* (rock) known today as Pyramid Rock. To Kānaka Maoli, however, she is the daughter of the husband and wife gods Kū and Hina, and is believed to have given birth to other pōhaku, which grew into larger rocks and helped to keep the Mōkapu Peninsula from eroding (Pukui, Elbert, and Mookini 1974, 119). Crossing into the ocean reconnects this surfer to her genealogy: grandmothers who surfed, great-great-grandfathers, and *nā akua* (gods: ke akua is singular, god, "nā" pluralizes akua to means "gods") who interacted with and wrote their history in the sea.

Diving under a wave, she feels the wet contact of water on skin, goose bumps rise in the rawness of the morning ocean, and her heartbeat accelerates as salt flavors her tongue. Placing her body in the sea, the Kanaka surfer is (re)articulating an indigenous cultural connection and claim to place by (re)enacting the sensory memories within the performance of heʻe nalu. When she hits the crest of a wave, her lungs puff in the same way that her ancestors' did. She thinks of the mele of Honokaʻupu, which tells of the moʻolelo of Māmala, a famous female surfer and prominent Oʻahu chief who had the *kino lau* (many body forms) of a gigantic *moʻo* (lizard) or a great shark. The mele speaks of a well-known surf break called Kalehuawehe that rose on Oʻahu and is now known as Waikīkī. Near Kalehuawehe, in front of what is now Honolulu, was a break called ke kai o Māmala, the sea of Māmala: "It broke through a narrow entrance to the harbor straight out from a beautiful coconut grove called Honokaʻupu and provided some of the finest waves in Kou (an early name for Honolulu)" (Finney and Houston 1996, 33). The break was named after Māmala. The *Mele of Honokaʻupu*:

> The surf rises at Koolau,
> Blowing the waves into mist,
> Into little drops,
> Spray falling along the inner harbor.
> There is my husband Ouha,
> There is the shaking sea, the running sea of Kou,
> The crab-like moving sea of Kou.
> Prepare the awa to drink, the crab to eat.
> The small konane board is at Hono-kau-pu.
> My friend on the highest point of the surf.
> This is a good surf for us.
> My love has gone away.
> Smooth is the floor of Kou,
> Fine is the breeze from the mountains.
> I wait for you to return,
> The games are prepared,
> Pa-poko, pa-loa, pa-lele,
> Leap away to Tahiti
> By the path to Nuumehalani (home of the gods,)
> Will that lover (Ouha) return?
> I belong to Hono-kau-pu,
> From the top of the tossing surf waves.

The eyes of the day and the night are forgotten.
Kou has the large konane board.
This is the day, and to-night
The eyes meet at Kou.[1]
(Westervelt 1915, 54)

The effervescent swells forming on the horizon, the slapping of the textured surface against the Kanaka surfer's knees as she sits atop her board, and the wafts of seaweed in the wind stir memories filled with history and knowledge. She is immersed in an indigenous epistemological awareness as her eyes feel the historic energy of ke kai sung of in the mele of Honokaʻupu. Her connection to ʻāina and ancestors is reaffirmed by her practice of heʻe nalu, and so is her ability to autonomously construct a multisited identity as both indigenous and modern. She is no longer merely a "career woman," a "mother," or a "colonized" body, because to plunge into the sea is to plummet herself into a process of re-creation. She reenvisions times and spaces as Hawaiian times and spaces, which are independent from colonial times, spaces, and places. Surfing beside Kūʻau, the Kanaka surfer is "reading" her cultural knowledge and history written in the seascape as a contemporary individual.

The Kanaka surfer "reads" the visual text of ke kai with her kino. Her oceanic literacy, her ability to read and pronounce the many knowledges written in ke kai by her ancestors, is what enables the Kanaka surfer to embrace seascape epistemology. Oceanic literacy is the applied knowledge within seascape epistemology; it is the literacy by which seascape epistemology becomes relevant and empowered. To understand the concept of seascape epistemology, then, one must first understand what the embodied knowledge within it looks like, smells like, feels like, and sounds like, which I explore through descriptions of heʻe nalu, hoʻokele, and lawaiʻa, as well as through analyses of language, moʻolelo, poetry, artwork, and contemporary oral histories.

Oceanic literacy is a knowledge of the exact location on the outer reef where the surf breaks and where one's footing may still be found, it is a knowledge of the place in the sea where the water is black from depth or from deep holes in the rock, and it is a knowledge of where the sea is very shallow and impassable for canoes. Oceanic literacy is the ability to read the cloud colors that indicate a passing squall, or the ripples on the water's surface telling of an approaching gust or of a school of fish being chased by a larger predator. An oceanic literacy requires an oceanic sensibility.

There are rhythms involved in an oceanic literacy. The rhythms of the waves, the moon, the sun, the tides, the fish, the winds, and the birds tell Pacific Islanders of the spawning seasons and times, when the waves will be good for surfing, when the ocean will be calm and clear for fishing, when the winds will be good for sailing, and where "home" is in relation to a destination island ahead. The rhythms of the clouds and currents communicate information, as living guides gliding along, within, and beyond the laws of the universe, requiring both an intellectual and a spiritual reading.

There is also a biological element involved in oceanic literacy rooted in certain facts that affect our notions of space and place: the body can only lie prone or stand upright and turn left or right. When surfing, diving, or navigating, however, the body breaks through these limitations. Surfers glide and float on water, sliding on a fluid floor, expanding the realm of gravity by vertically flowing up and then down, and sometimes racing through a stationary tunnel of pitching water. A navigator's sight, body, and gravity are also expanded as she floats upon a moving mass of liquid, relying on the ever-moving sky, birds, and winds to guide her toward a destination seen only in her mind. Diving to spear fish places the body upside-down, horizontally and diagonally aligned with the swaying seaweed as one's lungs compress between the bubbles. Spatial conceptions are altered, as is one's sense of place when engaged in an oceanic literacy, because of the ways in which one interacts kinesthetically with the ocean, the ways in which one physically involves oneself with the sea.

"Experience" is a term for the various modes through which a person knows and constructs reality. These modes include all of the senses, as a navigator inhales the salty air on a wa'a and a surfer tastes the blue-green liquid when diving under a wave. The element of emotion is also crucial in one's engagement in an oceanic literacy: fear, elation, tranquility, anticipation, anxiety, and pride as a surfer races down the face of a steep, quickly breaking wave with a shallow reef below, as a fisher hooks a large *ulua* (giant jack trevally), or as a navigator follows the silent footsteps of the stars across the sky.

Yi-Fu Tuan, retired professor from the University of Wisconsin, has pioneered the field of human geography, merging it with philosophy, art, psychology, and religion to explore how humans inhabit and shape the earth, not just with their bodies but with their emotions, minds, and spirits. In his work on space and place, Tuan notes that "it is a common tendency to regard feeling and thought as opposed" (Tuan 1977, 10). In oceanic literacy, however, they are both necessary and work together simultaneously. Emotion and logic are both ways of knowing. A surfer knows that when the ocean's surface churns violently with the force of a large river, she must not fight the current, or else tire and possibly drown. Instead, she must allow the feeling of fear to immobilize her temporarily and let herself be carried by the ocean until the current releases.

In looking specifically at the oceanic literacy of hoʻokele, it must be noted that much of this knowledge is learned through experience and intuition. Tuan explains, "He [the navigator] learns to detect reefs from the subtle changes in the color of the water, and he learns to read the sky" (81). Kanaka master navigator Charles Nainoa Thompson says that within a day, a navigator observes about three thousand different elements from which he or she must make about two hundred decisions (Thompson 2008). It is the integral experience rather than the deliberate calculation that informs the many decisions the navigator must make in the course of a long voyage. Tuan continues, "A navigator needs keen eyes, but he must train the other senses to a high degree of acuity as well," sometimes excluding visual cues to concentrate on other signs, as the stars may not be visible, and the wave patterns can be difficult to interpret visually from the level of the canoe (Tuan 1977, 82). One accomplished navigator from the Society Islands, Tewake, claimed that he would sometimes retire to the hut on his canoe's outrigger platform, where he could lie down and more accurately direct the helmsman to the proper course by analyzing the roll and pitch of the canoe as it danced in a distinct rhythm over the waves (Lewis 1994).

As a navigator, it is critical to expand one's concept of "space" (this is a concept further developed in chapter 3). Basic movements are key to the awareness of space. "Spatial ability" is the ability to walk, crawl, find one's mouth, and so on. "Spatial knowledge" is different, and takes more time to develop, and can later exceed what the body can do in ability. Spatial ability becomes spatial knowledge when movements and changes of location can be envisaged. It is about image-making power, which every navigator must embrace.

Tuan notes, "Many animals have spatial skills far exceeding those of man; birds that make transcontinental migrations are an outstanding example.... Mental worlds are refined out of sensory and kinesthetic experiences. Spatial knowledge enhances spatial ability" (Tuan 1977, 74). Expanding spatial knowledge to include the stars, the deep corals, and transient winds allows for movement within an expanded space, for communication with an expanded world. The realm of knowledge comes to include the knowledge of stones, birds, sharks, rains, planets, and gravitational pull.

In these ways, oceanic literacy is an alternative political and ethical literacy for Kānaka Maoli, offering movement and flexibility within a (neo)colonial system that continues to marginalize, compartmentalize, and subjugate. Oceanic literacy becomes a poiesis of nonlinear movement between spaces and times that creates a specifically Kanaka worldview and epistemology. The rhythms of this poiesis are reflected in the wrinkling of the skin and the flicking of the tongue, in the kinesthetic tempos that follow the music of the sea, its "language": morphology, syntax, and phonology. Hearing the sea's language through an oceanic literacy engages a process of contextualizing textures, colors, and moods through a political aesthetic of place.

For instance, in ʻŌlelo Hawaiʻi, *kai* is

calm, quiet sea, *kai mālie, kai malino, kai malolo, kaihoʻolulu, kai pū, kaiwahine,*
kai kalamania, kaiolohia
strong sea, *kai koʻo, kai kāne, kai nui, kai nuʻu, ʻōkaikai*
rough or raging sea, *kai pupule, kai puʻeone, kai akua, ʻōkaikai*
deep sea, *kai hohonu, kai ʻau, kai hoʻēʻe, kai lū heʻe*
restless sea with undercurrent, *kai kuolu, kai holo, kai lewa, lapa kai, kai kō, kai*
au
dark Blue sea, *moana uli, moauli*
streaked sea, associated with Kona, *kai māʻokiʻoki.*
whispering sea, associated with Kawaihae, *kai hāwanawana*
salt sea, *kai paʻakai*
shallow or reef sea, *kai kohola, kai koʻele*
rippled or ebbing sea, *kai heʻe, kai emi, kai mimiki, kai hoʻi, kai nuʻu*
aku
running sea or current, *kai holo*
western sea, *kai lalo*
high sea, *kai piha, kai nuʻu*
of the sea, *o kai*
place where sea and land meet, *ʻae kai,* By the sea, *a kai*

sea almost surrounded by land, *kai hāloko.*

the eight seas (seas about the Hawaiian Islands, poetic), *nā kai ʻewalu*
(Pukui and Elbert 1986, 114 *Hawaiian Dictionary*)

Each condition and locational niche within ke kai is recognized. Samuel M. Kamakau tells us, "*Ka poʻe kahiko* [the people of old] distinguished by name the waters along the coast, out to sea, and to the deep ocean" (Kamakau 1976, 11). For example:

opening, a calm place in the high sea or deep inside a shoal, *kīpuka*
rising of water when wind and current meet, *ʻakūkū*
ripple on the water, the rising up of water from the wind, *hāluʻa*
foamy sea, as when wind and current are contrary, *kai ahulu.*
place where pointed clouds or clusters of them arise out of the
    ocean, *pōpuakiʻi*
any place elevated in the manner of a high path, bank formed by
    sand at the
mouth of a river, *āhua.*
the current in the ocean, *au*
turn of the tide, *nioke*
bare reef, reef flats, shallow place of water some distance from the
    shore like
Kālia on Oʻahu and as at Kona, Molokaʻi, *kohola*
place far out at sea where fish are caught with a hook, *aukaka*
chopping sea, sea beyond the *poʻana*
second place where the surf breaks and where a footing may still be
    found, *ka helekū.*
place in the sea where the water is black from the depth or from
    deep holes in the rock, *puhi*
very shallow sea, unsuitable for canoeing in, a thumping sea,
    because the canoe thumps the coral, *kai koʻele.*
a place in the sea where a fisherman would look for *heʻe*, octopus,
    after blowing
chewed *kukui* nut in the water to enable him to see clearly (*ʻōkilo*),
    kai ʻōkilo heʻe
(Kent 1986, 253–58)

In ʻŌlelo Hawaiʻi, the diverse types and locations of beaches, coastlines, sands, corals, shells, marine life, waves, currents, tides, and salts within the sea were not "studied" but engaged. In an oceanic literacy, the flow of the mind matches the flow of the ocean. Reference and expression work together. ʻŌlelo Hawaiʻi mimics the lexicon of waves reflected in the matrix of aquatic molecules forming and breaking in an energetic hula:

> That which swells and rolls in "furrows" (ʻaui kawahawaha) just
> makai of the surf line (kuaʻau) is a nalu, a wave.
> A wave that breaks along its entire length is a kai palala, nalu palala,
> or lauloa; if it breaks on one side, that is a nalu muku.
>
> A wave that is sunken inward when breaking (poʻo poʻo iloko ke
> poʻi ana) is a nalu halehale (cavernous wave) [called a "tube" by
> modern surfers]; one that draws up high is a nalu puki; one that
> does not furrow or break is an ʻaio, a swell; one that sinks down
> just as it was about to break is a nalu ʻopuʻu.
>
> A wave that swirls and "eats away" [the sand] (poʻai ʻonaha) is a nalu ʻaʻai
> or ʻaeʻi; one that rolls in diagonally (waiho ʻaoʻao mai) is a nalu kahela.
> (Kamakau 1976, 12)

Singing about the movements of waves diving in and out of Hawaiian shores identifies specific seas, each holding specific purposes and potentials:

> Then comes the kai heʻe nalu, surf-riding sea, or kuaʻau, and the
> poʻina nalu, or poʻina, where the waves break.
>
> Just beyond this surf line is the area called kua nalu, back of the
> wave, or kulana, pitch and toss, and then the kai kea, white sea; or
> kai luʻu, sea for diving; or kai paeaea, sea for pole fishing.
>
> Outside of there are the areas of the kai ʻo leho and kai ʻokilo heʻe, sea
> for octopus fishing; the kai kaka uhu, sea for netting uhu; the kai
> kaʻili, sea for fishing with a hook and line; and the kai lawaiʻa, sea
> for [deep sea] fishing. (11)

ʻŌlelo Hawaiʻi translates words into an interactive literacy that instructs the Kanaka body in its actions, in its mobilized poiesis through performance. The seaʻs sounds and movements taught Kānaka Maoli how to physically move their tongues and fluctuate their throats when chanting their human histories. Winona Beamer, a contemporary kumu hula (hula instructor) tells how her grandmother, Helen Desha Beamer, taught ancient Hawaiian chanting and dance forms by taking her and her fellow hula students to the ocean and instructing them to observe the rise and fall of the surf, how to embody the ocean's rhythm by mimicking its ebb and flow:

> She had us stand on top of huge boulders by the ocean and practice chanting over the sound of the waves. The sea breeze, sometimes a wind, rushed in. The surf, turning paisley where it rolled in over the reef, bounded in pure energy toward us. As the waves grew and came nearer we stood firmly and chanted louder, louder—until the waves broke in fantastic turquoise and white plumes. Even as we got doused, we chanted. And as the water ran back to the sea, we chanted, we chanted more softly. Often our names were chanted.
>
> Sometimes a wave would curl on the beach, then make delicate ripples before it slipped back into the depths. As it moved away, we were supposed to hold a quavering trill until another set of waves started toward us. (Lueras 1984, 34)

Kānaka Maoli articulate the sea's breath and song through the sensorimotor skills of the throat and tongue, recognizing how these rhythms embrace both repetition *and* change. The sun will always rise over the reef at Makapu'u Point on O'ahu, but never over the same wave, nor with the same intensity or brilliance as the day before.

The tone and vocabulary of 'Ōlelo Hawai'i reflect the ocean's many movements, which speaks to an epistemology about how the world functions and exists. One critical element inherent in this oceanic epistemology is the connection, rather than the separation, of places. Kanaka waterman and author John R. Kukeakalani Clark explains,

> In the western concept, there's a real dividing line between land and sea. . . . The Hawaiians were just the opposite. To the Hawaiians, the ocean is just an extension of the land, it just happens to have some water on it. It's an extension of the land, the reef out there is just as valid a farming area if we're talking subsistence. It's as a lo'i [taro patch] is on land for taro. So you'll find that all the reefs are named, all the fish holes are named, there's names that you find for plants, fish, crustaceans in the ocean that are the same as the names on land. So they're not really differentiating; it's all one. It's all 'āina. That's [pointing to the sea] just 'āina with water on it. (Clark 2007)

The language of oceanic literacy joins the world together (Carter 2009). Building and naming waʻa, for example, involves a literacy that embraces the waʻa as a manifestation of the forest, carrying the mana of the trees into the sea. Bruce Blankenfeld explains,

> The canoes are named with one purpose, to acknowledge the spiritual nature of this thing that was built that was once a living thing in the forest, and now it is part of the ocean. But in an island society there is no difference between the land and the ocean. They are one, they mirror each other, and through that idea, everything on the ocean that is named, all the fish, all the limu, all the types of coral and everything, has a counterpart on the land. . . . Hawaiians figured out a long time ago, with the ahupuaʻa system, that everything that comes down from the land effects the ocean and there's a duality there. They need each other. Just like in life, a man needs a woman. So when a canoe goes in, you have a whole way of addressing this spiritual nature of it, there's a naming process, and it [the waʻa] tastes the salt water and now it belongs, its spirit and everything belongs to the sea. (Blankenfeld 2008)

Emphasized in Blankenfeld's oceanic literacy of building waʻa are perceptions of interdependence, interconnection, and a cyclical approach to the way the world functions. In an oceanic literacy, places become fluid as opposed to divided, and full of historical significance instead of perceived as empty and thus available for consumption. Revealed is how Hawaiian places are filled with gods and ancestors residing in the ʻāina as manifestations of both terrestrial and aquatic beings. Kanaloa, for instance, has many kino lau found both in water and on land: heʻe (octopus), koholā (whale), honu (sea turtle), maiʻa (banana), ʻalaʻala pūloa (small weed), and the island of Kahoʻolawe itself. Na akua (gods), ʻaumākua (family or personal gods), and na kupua (demigods), have land-ocean kino lau in human, animal, plant, or mineral form, as well as manifestations of actions and meteorological phenomena. Kamapuaʻa, for instance, a kupua associated with regeneration and fertility, is able to adapt; on land he takes the body of a pig, and in the ocean, he can become a humuhumunukunukuāpuaʻa (trigger fish).

While oceanic literacy is not a written literacy in Western notions, remaining outside dominant systems, it is written by Kanaka ancestors in an organic language, in the sea's reefs and fishing holes. History can be read in the first fishpond in Hawaiʻi, which is said to have been built by the fish god Kūʻulakai at Lehoʻula, ʻAleamai, in Hāna, Maui, where he lived with his wife

Hina and their son Aʻiaʻi. Beside this fishpond lies a stone that is the body of a chief:

> His marvelous fishpond attracted a lot of attention especially because the pond, through his power, was always full of fish. At Wailau on Molokaʻi was a chief who had the power, as a kupua or demigod, to turn into a gigantic eel, 300 feet long. In his eel form he was attracted to the fishpond and slipped into the inlet. However, after he had fed well he could not get out without breaking down the wall. He hid in a deep hole beyond ʻĀlau island called "hole of the ulua," and Kūʻulakai baited the famous hook, Manaiakalani, with roasted coconut meat to lure him out of hiding. When hooked, the eel was dragged ashore by two ropes held by men who stood on opposite sides of the bay while Kūʻulakai stoned the eel to death. The body of the eel turned to stone and can still be seen today. (McGregor 2007, 89)

Na koʻa (altars) used by Aʻiaʻi to mark fishing grounds by and in honor of Kūʻulakai and Hina, as patrons of fishing, are also part of Hawaiian oceanic literacy: "The first fishing ground marked out by Aiai is that of the Hole-of-the-ulua where the great eel hid. A second lies between Hamoa and Haneoo in Hana, where fish are caught by letting down baskets into the sea. The third is Koa-uli in the deep sea. A fourth is the famous akule fishing ground at Wana-ula. . . . At Honomaele he places three pebbles and they form a ridge where aweoweo fish gather. At Waiohue he sets up on a rocky islet the stone Paka to attract fish" (Beckwith 1970, 22).

These koʻa are also historic sites of ocean knowledge and indicate which types of native fish would naturally conglomerate along certain sections of deep sea mounds, and how they were cultivated for sustainable fishing. Kānaka Maoli would historically gather seaweed as feed that would be taken out to these koʻa in canoes, ensuring that the fish remained available and plentiful in this designated area. Today, Kānaka continue to practice this oceanic literacy by caring for the sites. Walter Paulo and Eddie Kaʻanana, two fishermen from Miloliʻi on the island of Hawaiʻi, know and *mālama* (care for) the koʻa in Miloliʻi, which certain families still access for fish today. The significance of knowing the locations of as well as ways of caring for these koʻa continues to act as a critical literacy for Kānaka Maoli, particularly in more remote and rural areas such as Miloliʻi that sit outside dominant narratives that determine how successful educational and economic systems should function in a modern society.

The pervasiveness of the fish god Aʻiaʻi in Hawaiian literacy permeates all of the islands. What is known today as Tongg's at Diamond Head on Oʻahu, a very popular surfing spot for locals and commercial surf camps, was historically named Kaluahole Beach by Kānaka Maoli, after being visited by Aʻiaʻi (Clark spells the name ʻAiʻai). Clark narrates the moʻolelo of this surf spot written in the caverns and rocks of the seascape: "ʻAiʻai, the son of Kūʻula-kai, the Hawaiian god of fishermen, was commanded by his father to travel through the islands of Hawaiʻi and teach the people all the ways of fishing and to establish fishing stations and shrines along the islands' shorelines. After ʻAiʻai established a station at Kāhalaiʻa on Oʻahu, he traveled onto Kaʻalāwai, where he placed a brown and white rock in the ocean. In this rock was a cavern filled with the *āholehole* fish, so it was appropriately named Kaluahole, the *āhole* cavern" (Clark 1990, 43).

Bringing ʻŌlelo Hawaiʻi and moʻolelo about ke kai back through oceanic literacy helps to remember the spatial histories of Kānaka Maoli; the ontological, epistemological, economic, spiritual, and personal relationships and responsibilities to ʻāina that are critical to Hawaiian identity and the political economy. Kanaka identity is, after all, physically and metaphorically borne up by the ʻāina. Moʻolelo tell how the land and sea are often literal body parts of Kanaka gods. One moʻolelo tells of how Sandy Beach on Oʻahu is the location of an actual anatomical part of Pele's sister's body: "In the legends told of Pele, it was said that a sister of the fire goddess, attempting to distract Kamapuaʻa, a handsome demigod, from his pursuit of Pele, threw her own vagina to this spot, where he followed it. Kohelepelepe, meaning labia minor, refers (with Hawaiian directness) to the resemblance between the natural shape of the crater (when viewed from a distance) and this part of the female anatomy" (25).

Kanaka scholar Davianna Pōmaikaʻi McGregor narrates another moʻolelo of how the Hāna Coast on Maui was formed from the body of Pelehonuamea, or Pele: "The goddess dwelt at Haleakalā and built it up to its present size until her mortal enemy—her sister, Nāmakaokahaʻi, an ocean deity who could assume the form of a dragon—discovered where she lived. Nāmakaokahaʻi arrived at Haleakalā with another sea dragon, Haui, and together they viciously attacked Pelehonuamea and dismembered her body. Parts of the body landed in Hāna near Kauʻiki and formed the hill called Kaiwiopele (the bones of Pele)" (McGregor 2007, 90).

Ka ʻāina is made from Kanaka ancestors; the genealogical relationship can be felt underfoot and with the hand. Kanaka blood has been mixed within the land and sea. At Kuamoʻokāne, which is known today as Koko Head, on Oʻahu,

> A Hawaiian legend tells us how the land there came to be red. A chief and chiefess of the nearby district of Waiʻalae had a daughter whom they gave up in adoption. When the girl reached maturity, she went one day to the home of her real parents to visit with them. They were not there when she arrived. While waiting for them to return, she picked a stalk of sugarcane and ate it. Then she went out to the point now called Koko Head and swam in the sea. The unfortunate maiden did not realize that her parents had a shark god whose duty was to kill anyone who molested the food-plants they had cultivated. While she was swimming, the shark attacked her, and the blood from her wounds spurted upon the land. From that time on, this place was called *koko*, or "blood." (32)

Kanaka scholar and poet Haunani Kay-Trask sees the rich purple and red Koʻolau Mountains on Oʻahu as running with the blood in her veins. Her kino reflects the tones of the ʻāina, born up from the taro in the dirt of her native island. Trask writes in her poem, "Koʻolauloa":

his earth glows the color
of my skin sunburnt
natives didn't fly

. . .

of this ʻāina
their ancient name
is kept my *piko*
safely sleeps

. . .

I know these hills
my lovers chant them
late at night

owls swoop
to touch me:

ʻaumākua
(Trask 1994, 80)

Kānaka Maoli are like marinated souls steeped in a liquid filled with millions of years of knowledge and memories. Flesh and bones are written into the shapes and movements of ka 'āina. This intimate interaction and relationship can inspire the contemporary mind, through an embodied experience, to imagine this past, including what has been lost, for a reconstruction of the future. Oceanic literacy becomes a political language for Kānaka Maoli.

Politics of Oceanic Literacy

'Ōlelo Hawai'i was increasingly disparaged in the mid-nineteenth century following the 1820 arrival of missionaries, who determined 'Ōlelo Hawai'i to be an inadequate tool for "progress." The teaching of 'Ōlelo Hawai'i was officially outlawed in public and private schools after annexation (military occupation) in 1893, and at this time English became the only acceptable language for business and government in Hawai'i (Silva 2004, 3). This attempted linguistic effacement (I say "attempted" because 'Ōlelo Hawai'i continued to be spoken, chanted, and written from the time of the arrival of missionaries through statehood and up until today) that accompanied colonization profoundly affected and continues to affect Hawaiian knowledges. Feminist scholar Ramona Fernandez explains, "The imperfect destruction of other knowledges and the inevitable multicultural exchange resulting from that imperfect destruction are repressed in the Western history of knowledge. But rememory is at work recovering and reinventing these knowledges" (Fernandez 2001, 12). Kanaka oceanic literacy is one way in which contemporary Kānaka Maoli are recovering and reinventing, mobilizing codes of reading, writing, and reorganizing the world in ways that are most beneficial to Kānaka (12).

Oceanic literacy, which is part of 'Ōlelo Hawai'i, interrupts what Jacques Rancière calls the "distribution of the sensible by supplementing it with those who have no part in the perceptual coordinates of the community, thereby modifying the very aesthetico-political field of possibility" (Rancière 2004, 3). Rancière's concept of the "distribution of the sensible" illuminates the potential that oceanic literacy has to make Kanaka voices audible by expanding the horizon of the "sayable." In this way, a Kanaka surfer becomes more than merely a body riding a wave; she can also become political through the sensibility of the act that represents a historical Kanaka way of knowing. It is not only the enactment but the imagination of riding the raw power of a wave, even without the experiential component of the literacy, that makes visible new spaces for all Kānaka Maoli. This is the potential of oceanic literacy as an aesthetic political literacy; this knowledge has the ability to engage all types of movement. And the body sits inside this movement, repositioned by the sea into physical and philosophical states that are more receptive to ways of being-in-common and thinking-in-common with others.

Within a Hawaiian context, oceanic literacy becomes an aesthetic logic that remembers through performance. The movements of the body inter-act with language, reading, and writing so that the literacy does not merely employ the eyes, brain, and fingers but also a kinesthetic engagement with one's surroundings. It is both reading and writing through empirical obser-vation as well as subjective sensations. Oceanic literacy remembers what was written in the coral and on the fins of turtles through an active interac-tion. This embodied literacy splashes mobile trails into the sea, inviting a re-sponse from the sea itself. The ocean is involved in the writing and reading process, affecting how we create and shape both our selves and our nations.

Reaffirming identity through relationship to place is not an act of ideo-logical purism, as seen in the rigid and absolute boundaries of religious fundamentalism, for instance, where there is no "space" for negotiation or fluctuation. In an oceanic literacy, the sand always shifts, and this affects surfing lineups and fishing holes. The stars, sun, and moon are always cir-culating, presenting patterns but never absolutes. While all constructions of place seek to impose a structure, some way of ordering the world, seascape epistemology resists imposing a specific framework, allowing for individual interpretation and adjustment within the continually changing and growing space and time of the ocean. The seascape's boundaries are never complete, and thus a state of change is inherent within the epistemology, because it *is* change.

Nor is an engagement of oceanic literacy the affirmation of an authentic knowledge, because while it asserts cultural sovereignty, oceanic literacy also invents something new for Kānaka Maoli that is continually (re)created with the modernization and development of Hawaiian bodies, minds, land, and sea. The performance of oceanic literacy is an experience with the ʻāina that does not presume a specific, "indigenous" way of being-in-the-world revealed through the knowledge. Instead, oceanic literacy offers a means of constructing a fluid identity anchored in place. An indigenous identity is a subject in process as opposed to one in stasis, or one that is already complete.

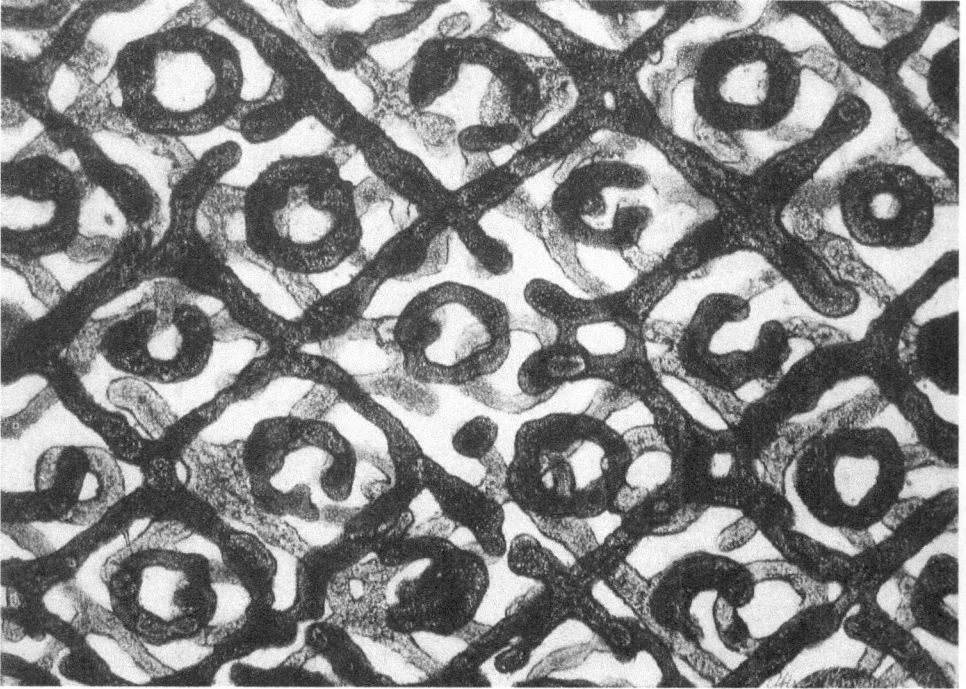

FIGURE 2.1. *Maka Upena* (Watermark), 2008. Abigail Lee Kahilikia Romanchak. *Kapa* (bark cloth) print. Courtesy of Abigail Lee Kahilikia Romanchak.

Kanaka artist Abigail Lee Kahilikia Romanchak illustrates this fluid articulation in her indigenous interpretation of what is predominantly perceived as a static symbol of lines and boundaries: a net. In her *kapa* (bark cloth) print *Maka Upena* (see figure 2.1), Romanchak creates a contemporary Hawaiian watermark that replicates patterns found on the surface of the sea as well as the pattern of a Hawaiian fishnet, a historical and modern tool used for gathering food.

The display of circles and lines do not create grids, as might be expected in a net. There is connection, illuminating the interrelationships between Kānaka and ke kai, but the net forms organic shapes and pathways between the lines of the net. Some lines become circles, and circles become lines. There are layers and different hues and depths of blue, as there are in the ocean. Romanchak's kapa print reinforces the Kanaka conception of the sea as a meandering melody, whimsically singing in variant pitches and tones.

Imagine a Kanaka fisher using a real fishing net, potentially blurring lines in another way. This fisher might not be able to write her name on paper, but she draws an ethical line in the ʻāina between herself and a blue-green *uhu* (parrot fish) she hunts. The fisher knows from her oceanic literacy that this blue uhu is a male, as opposed to the female uhu that are red. This fisher also knows that if she were to take this blue fish, she would remove the only male from the school, and the harem which uhu form will not be able to reproduce for another year, until one of the red female uhu finally transforms into a male. This fisher blurs conventional boundaries between literate and illiterate. Knowledge of the sea breaks from an assimilated system and spills out, dis- and re-organizing definitions of literacy, as well as politics and ethics.

Oceanic literacy is a critical political and ethical literacy in today's era of global warming that has resulted in the mass destruction of coral reefs due to acidification and bleaching, declining fish populations, and other negative changes in the world's marine ecosystem. Many Pacific Islanders living on low-lying atolls or islands, while responsible for very little of the increasing amounts of greenhouse gas emissions trapped in the atmosphere, are now threatened by rising ocean waters around them. The documentary *Rising Waters: Global Warming and the Fate of the Pacific Islands*, produced by Andrea Torrice in 2000, reveals how unusually high tides have swept the low-lying atolls of Micronesia, destroying crops and polluting fresh water supplies. The film also speaks to how the cultures of seven million inhabitants across the Oceanic nations of Sāmoa, Hawaiʻi, the Marshall Islands, Kiribati, and others are also being threatened as waters rise. Ancestral sites and graves have been destroyed due to increased erosion and from an increase in the frequency and intensity of hurricanes, both phenomenas directly related to global warming. It takes only a few feet of rise in the level of water to destroy some of these Pacific nations. Meanwhile, those Western nations most responsible for emitting the greatest amounts of greenhouse gases are laboriously tying their hands with debates on how much and how to lower levels of future emissions (if at all). Within this capitalist debate, the disconnect between person, nation, and seascape is profound. The industrial world remains preoccupied with the immediate and potentially negative economic implications rather than with the long-term economic, political, social, cultural, and human costs to their outdated practices.

Oceanic literacy becomes an ethical reading of the ocean that exists outside these dominant political and economic interests. The performance of oceanic literacy brings human awareness back into the sea, integrating not only the conscious mind but also the physical body back into the interactive processes of transformation continually occurring between ʻāina and kino. Human and environmental bodies affect each other, creating a complex whole from distinct and incoherent parts. Oceanic literacy helps to remember the historical consciousness within place and our human moral connection to the planet. Without this interaction, Paul Carter warns, "We proceed without memory as if the spaces we inhabit are a *tabula rasa* that we can choose to inscribe as we wish. The human environmental and spiritual costs of this collective forgetfulness are everywhere to be seen, in the reckless destruction of cultures, in the overexploitation of the earth's gifts, and in the delirium that passes for free choice in our consumerist society" (Carter 2009, 84).

Ethics of Oceanic Literacy

"When I introduce myself, I am born of the same gods that gave birth to these islands, gave birth to the ocean, to the things that crawl in the sea, that crawl on the land. Being the last born, that's my relationship to all, everything that came before me. As the last form, it becomes your kuleana, the responsibility to care for all that came before you . . . take care of the ocean which gives life to the land, which in turn gives life to the ocean" (Holt-Takamine 2007). Kumu hula Vicky Holt-Takamine places her identity within ʻāina so that she does not inhabit a "tabula rasa." She does not "know" the ocean in an isolated manner, but relates to it through a genealogical connection of what and who came before. This knowledge belongs not to her but to her culture—a perception of knowledge that permeates seascape epistemology and establishes a different way of using knowledge not anchored in human "progress" or "enlightenment," as advocated by the Industrial Revolution, but as a recognition and acknowledgment of human responsibility to all forms of life. A redirected definition of progress can reorient an ontology of the world not subject to a mastery and domination of the environment by establishing intimate relationships and by helping to embody fluid connections between the self and the natural world.

Jane Bennett advocates for this ontology as an alternative to what she argues is a dialectical debate currently dominating the American environmental discourse. The debate wrestles between two perspectives that reflect Hegel's dialectic between faith and enlightenment: "nature holism," a cooperative approach to the environment in which human beings have a symbolic place in what is perceived as an enchanted and providential natural world, and "environmental management," which views the environment as a problem susceptible to rational control (Bennett 1994). Bennett asserts that within this struggle, each opposing perspective depends upon the other for its position, constraining thought within a singular ideological framework and predisposing both to specific ethical outcomes.

For instance, in natural holism, the belief is that the order of and meaning within nature has been created by God (or some divine form). Yet within this belief, in which order and meaning come from God, the human understanding of this order and meaning remain incomplete and inaccessible. The result is an ambiguity and uncertainty about our human place within nature that leads to skepticism and distance. Environmental management, in contrast, assumes the path of enlightenment, hoping to reduce the ambiguities of nature through a rational measuring, controlling, and ordering of wilderness. The epistemological and ethical positions of both views are strictly limited by the dialectical framework in which they are mutually defined: natural holism defines its views in relation to its opponent and is thus unable to envision a secularized orientation to nature, while environmental management cannot envision a respect for nature that does not recall an enchanted world (Bennett 1994). Bennett argues that the ethics derived from both dominant narratives are unable to lead us toward a nondestructive relationship with nature because nature is never respected as Other. What she calls for is the imagination of another ontology.

Seascape epistemology offers such an alternative by recognizing the interconnection between ke kino and ke kai, a connection that, in part, constitutes their differences. Movement in one affects the state of the other. The sea is Other, but the body and mind become part of it, not in totality, because the ocean can never be fully understood by human capacity, but, as Hegel recognizes, by connecting as irreconcilably opposing concepts. Seascape epistemology reorients our relationship to the environment by breaking with this dichotomy: a continuum of domination and the enchantment of complete harmony. Holt-Takamine connects to ke kai through her genealogical lineage to the beginning of space and time by placing herself within this lineage not as something entirely coherent but as part of the complex whole (Holt-Takamine 2007). Big fish eat little fish, and hurricanes uproot coral beds; the seascape functions as multiple (and at times violent) parts of one whole. An ethics evolves that accepts and tolerates nature's unpredictability as part of one's experience in this world.

As the ocean is a powerful body that human beings cannot control, engaging it requires an interaction with its present form and mood, both of which are never stable. Such is the experience of life. The seascape exemplifies the significance and benefit of establishing relationships anchored in respect and an understanding of limitations, from which ethics grows: cooperation, adaptation, and humility. Seascape epistemology embraces the ambiguity and fluctuation within the ocean rather than being skeptical of or distanced from it. Seascape epistemology approaches difference as an interactive relationship rather than a rigid dichotomy. Preserved is a concern for Otherness, a relationship of respected alterity that forges rich, complex, and, paradoxically, intertwined identities.

An example of this kind of ethics is offered in the sixty-two-foot Hawaiian *wa'a kaulua* (double-hulled canoe) *Hōkūle'a*, first conceived in 1973 in the imaginations of Kanaka watermen Herb Kawainui Kāne and Tommy Holmes, and anthropologist Ben Finney. From the moment of conceptualization through the present day, *Hōkūle'a* has been a renowned symbol of Kanaka pride, empowerment, and mobility. The 1970s witnessed a revitalization movement in indigenous knowledge, not only in Hawai'i but throughout Moana. New maps were being drawn based on the idioms created by Oceanic peoples, and these idioms imagined new paths across the Pacific. These new maps were born within this new imagination with the specific intention of proving that Pacific Islanders were able to travel across great distances over eight hundres years ago, purposefully settling the islands of the Polynesian Triangle, and then finding their way back home again using only oceanic navigational concepts and methods.[2] In 1976, *Hōkūle'a*'s first deep-sea voyage from Hawai'i to Tahiti succeeded. This first voyage of the Polynesian canoe, however, had to be navigated by a master navigator from the island nation of Satawal in Micronesia, Mau Piailug. Because the oceanic literacy of ho'okele had been lost in Hawai'i since colonization, Piailug not only enabled the success of this first voyage but also the larger reawakening of voyaging knowledge in Hawai'i. Piailug consistently mentored and supported Kānaka Maoli in the relearning of a contemporary Kanaka version of ho'okele. Piailug enabled a cultural renaissance that began in the 1970s in Hawai'i that has grown into a significant cultural movement in support of Hawaiian values, knowledge, and dignity.

As an act of gratitude and *aloha* (love, respect, appreciation) for this priceless gift, in 2007, *Hōkūleʻa* embarked on a journey from Hawaiʻi to Satawal called Kū Holo Mau, "Sail On, Sail Always, Sail Forever." On this voyage, *Hōkūleʻa* escorted the voyaging canoe, *Alingano Maisu*, to Satawal, offering the canoe as a gift back to Piailug. The voyage was an ethical movement meant to honor Piailug as an individual, the knowledge he imparted to Kānaka Maoli, and his expression of Pacific Island interdependence and inclusion. Along the route to Satawal, *Hōkūleʻa* and *Alingano Maisu* stopped at the Marshall Islands and Chuuk, physically touching these Oceanic neighbors with their message of Oceanic indigenous empowerment. These waʻa (re)drew cultural lines that (re)aligned Oceania, traveling across the Euro-American boundaries separating Polynesia, Micronesia, and Melanesia on a static map. The voyage was an event that reconnected distinct yet interconnected knowledges from Satawal to Hawaiʻi to the Marshalls, to Phonpei, to Chuuk, and back to Satawal, not merely as a process but as a means of engagement.

These waʻa have not invented an ethics, but have (re)set it in motion. The movements of *Hōkūleʻa* help to reinvent a people by arousing memories and making indigenous knowledge visible and tangible, even for those Kānaka Maoli and Pacific Islanders who have never set foot on a boat. The *Hōkūleʻa*'s arrival on the shores of Tahiti with Piailug in 1976 is an image of this reawakening of oceanic literacy in Oceania (see figure 2.2). It illuminates the power that the mere presence a waʻa kaulua has to pull people together.

The ethics of oceanic literacy travels like a wave: it is formed from the vertically deep Satawalese ocean of historical and cultural roots, and is shaped into a new and unique Kanaka potential when it hits the horizontal land. Upon contact, this ethical wave becomes linear and contemporary as the shores of the Hawaiian Islands modernize. Epistemology and ontology are reinterpreted by historic sources through a contemporary lens. Kanaka voyagers do not need to sail on an ancient waʻa kaulua to embody the essence and potential of the enactment; *Hōkūleʻa* was constructed with modern materials without losing its essence or its impact. It is the literacy within the act that constructs and creates a sensibility through spurts of agility and acumen steeped in an embodied awareness of place.

As Kānaka Maoli interact with the sea, its gestures take on a substance and a specific tenor that affects the body, which Henry David Thoreau believed can shape political wills, affiliations, and ideological commitments.

FIGURE 2.2. *Hōkūleʻa*'s arrival in Tahiti, 1976. © Nicholas Devore III/ National Geographic Creative.

"Wilderness" can affect and startle the body, even irritate it, into a new state of awareness, and Thoreau targets this affective register of ontology, asserting that senses and textures can free the body from a "typical pattern of anticipatory perception" (Bennett 1994, xxii). The self is "ejected" from banality, Bennett says, revealing the world's complexity and potential, a counterforce to the lure of conventional and dominant thought-worlds and forms of literacy. This politics, instigated by the natural world, becomes an ethics by providing an impetus to stimulate the body into an engagement with, a state of appreciation for, and a connection to both human and non-human dimensions.

Studying the stars and chanting the names of the winds places an importance on things emotional at sea. Exploring various conceptions of the clouds and currents, analyzing and coming to learn about marine life, and formulating values regarding human ways of interacting with the waves and seaweeds evolves a specific society. There is an aesthetic-affective attachment, an affective precondition of ethical generosity in social relations that resists the fundamentalism seen in dominant political-moral concerns that generate a reactive demand for certainty. The political and ethical self is not explained; it is enacted through an exchange. Oceanic literacy provides the tools that Kānaka Maoli can use for the (re)creation of an autonomous and multisited identity.

Getting to Seascape Epistemology

Sitting in the sea at Mōkapu, interacting with the pōhaku (rock) Kūʻāu, the Kanaka surfer is reminded of our human obligations to ke kai. The oceanic literacy of heʻe nalu, as well as hoʻokele and lawaiʻa, engages a conviction that the human and nonhuman worlds are united though movement that is at once physical and philosophical, rational and impassioned, political and ethical. These oceanic practices also (re)claim an indigenous relationship to ʻāina for Kānaka Maoli because they require a literacy steeped in historic native rights, knowledge, and genealogy, all engaged the moment a Kanaka surfer, navigator, or fisher dives into the time and space of the sea. Placing ke kino within Hawaiian spaces and times in the ʻāina is what enables Kānaka Maoli an alternative and autonomous way of existing within dominant thought-worlds.

If ke kino of the Kanaka surfer is juxtaposed with that of a surf tourist's body, for instance, also waiting to ride the same waves in the same sea at Mōkapu, a significant distinction arises. Both surfers feel the trade winds blowing with an onshore force from the ocean, hitting their faces with a sharp chill as they swing up the bounding cliffs of the Koʻolau mountains. Yet their distinct ontological and epistemological relationships to the wind and ʻāina unveil distinct possibilities for each surfer. The Kanaka surfer, attuned to her indigenous conception of ʻāina, listens for stories floating on the sea's breath. Each gust of wind tells her of connecting pathways between herself, the waves, and the beach; between the present, past, and future. The tourist surfer, foreign to the stories and history of this specific place, does not engage the wind in the same way. She sees and feels a time and space of ocean that is perhaps recreational, spiritual, or aesthetically soothing, but is unaware of the Hawaiian genealogy within the wet rocks. While the tourist surfer might embrace an ocean-based knowledge of surfing, and while the ocean might also shape her identity, her connection to this specific ocean can be acquired only through her own historical context.

This is significant because it reveals how the Kanaka surfer is engaging a way of knowing that is distinctly Hawaiian and thus offers her a specific form of political empowerment, an empowerment that becomes critical within (neo)colonial systems. This distinction is not meant to reenforce or create binaries between "native" and "nonnative." Identities are complex and overlapping. What the Kanaka surfer illuminates is how her specifically Hawaiian ontology emerges within the Hawaiian kai, and offers a means of distinguishing an indigenous epistemology about movement within and alongside other (sometimes dominant) epistemologies related to the ocean. Surfing beside Kūʻau enables an engagement with this Kanaka surfer's native ontology. The distinction is meant to include Kānaka rather than exclude the tourist. The distinction creates an indigenous space that is at once mobile and rooted, indigenous and modern, and it reaffirms the Kanaka relationship between kino and ʻāina, getting her to the realization of seascape epistemology.

SEASCAPE EPISTEMOLOGY   Ke Kino and Movement

The Dance of He'e Nalu

One day when the sea was unruffled by the wind, the famous waves of 'Uo began to break gently. Excited, the kama'āina [locals] hurried out to surf and Kūanu'uanu [the priest iwikuamo'o of Keawenuia'umi, the chief of the island of Hawai'i] joined them to show that this keiki [child] of Hawai'i was a skillful surfer. The ali'i [chiefs] of Maui were impressed by his surfing as he rode swiftly on the waves without getting wet; he wasn't overwhelmed by the surf as he rode the waves, breaking to the right, then to the left. His body looked magnificent as he stood up on the surfboard and rode toward shore. — NAKUINA 2005, 1–2

In *The Wind Gourd of La'amaomao*, Moses Kuaea Nakuina introduces he'e nalu as a multidimensional enactment—as an art, dance form, and beloved activity of the ali'i, holding political and social power. Nakuina goes on to describe he'e nalu as a noble and chiefly skill: "When he [Kūanu'uanu] knelt with his arms outstretched, he looked like a manu ka'upu [Laysan albatross] treading the surface of the sea. His fame as a surfer spread all over Maui" (Nakuina 2005, 2). Kūanu'uanu's skill in he'e nalu was coveted and won him respect among ali'i as well as *maka'āinana* (commoners) across the island chain. His son, Pāka'a, also became particularly literate in ocean-based knowledges, which won him land and honor: "He learned the laws of the skies and the nature of the earth; farming and all the activities related to it; astronomy and sailing the seas; navigating and steering a canoe; living in the uplands; and fishing and all the activities related to it. Because Pāka'a was so knowledgeable and skillful, Keawenuia'umi gave him a high position in the aloali'i [royal court] just under his father Kūanu'uanu" (20).

Oceanic literacies such as he'e nalu were respected by the ali'i because to surf across the face of a wave is an act of dignity, competence, and strength. To surf through the ocean on a *papa he'e nalu* (surfboard) or in a wa'a is to obtain a state of versatility and movement by using the energy of nature. He'e nalu is a dance expressing power, but not merely political or social power. He'e nalu is also a process of gaining spiritual power, or mana; a process of being still, through movement, by listening to your own heartbeat create a rhythm with the ocean's gestures. He'e nalu is an exchange through awareness and a form of meditation in which the mind exhales difference and the soul inhales an intimate bond steeped in time and culture. Sitting in the sea, upon a board, waiting for waves, the murmurings and shouts of the ocean are familiar to the Kanaka surfer, sounds that have taught her to sing and speak.

In *Ka Mo'olelo no Hi'iakaikapoliopele: The Epic Tale of Hi'iakaikapoliopele*, Ho'oulumāhiehie tells of the grace of the ali'i Punahoa as she surfed on the island of Hawai'i. Punahoa's body is linked to the *'iwa* (great frigate) bird, to the white crest of wave, and to an arching rainbow: "It was said that the gliding of the 'iwa bird on the fringes of the wind was only half as beautiful as the glory of this chiefess's stance when she skimmed the white crest of the waves. Her board soared out on the backwash, and came back in on a cresting wave. As she mounted the shoulder of the rising swell, a rainbow appeared over her, arching down her sides" (Ho'oulumāhiehie 2006, 77).

The mo'olelo speaks of how the goddess Hi'iaka and her traveling companions were also watching Punahoa, at which Hi'iaka commented to her companions, "The real skill in surfing is not letting the board be taken by the water. If the board is controlled by the wave, that's not expert surfing" (78). This is a fascinating description of the skill of he'e nalu, which does not attempt to control the ocean but to instead "not let the board be taken" by it. It is a game and an interaction that is intimate and humbling, one that offers moments of physical freedom from gravity by floating up, and others of quiet reflection. The oceanic literacy within he'e nalu supports the larger Hawaiian epistemological approach to life in which one's environment and oneself interact through a reciprocal exchange of give and take, control and submission. In he'e nalu there is never an attempt to dominate ke kai, but a surfer must be skilled enough to maneuver the mind and body through it. This relationship builds mana for Kanaka surfers; through he'e nalu, they slip into themselves as connected beings riding through a native sea. Bruce Blankenfeld explains, "Mana is a spiritual strength, but to me it really is something that is built through what you do, your interaction with the ocean. . . . As you go, you're stripping away all kinds of fears and failings which you spot, and you can deal with. That's where your mana comes from. It is to know who you are, that's mana" (Blankenfeld 2008).

Historically, mana was profusely sought by all Kānaka Maoli, particularly *mō'ī* (kings) and *ali'i nui* (high-ranking chiefs), for several reasons: it is divine power that protects and empowers; it helps one escape from the commonality of a laborer; and for a mō'ī to fail on the path to mana through one of two paths, through either Lono (ke akua of agriculture and the arts who celebrated human sexuality) or Kū (ke akua of war and political power), is to condemn oneself to being outside the state of *pono*, outside the state of being just or righteous. Being pono was critical because, in the words of Lilikalā Kame'eleihiwa, "When a *pono Mō'ī* was religiously devoted to the *Akua*, the whole society was *pono* and prospered" (Kame'eleihiwa 1992, 48).

Maka'āinana, however, could obtain mana only through sexual relations with an ali'i, who held an abundance of mana through birth as specific designation by the gods.[1] This means of obtaining mana for commoners follows the path of Lono, and it is by this path that low-ranking *kaukau ali'i* (a lessor chief) and maka'āinana could attract the attention of an ali'i nui through activities available to all social ranks, activities such as he'e nalu. Of the two paths to mana, only the path of Lono has survived today. Kame'eleihiwa states, "Modern Hawaiians celebrate life through sports, hula, and sexuality. Hawaiian youth devote themselves to preparation for the many hula festivals and canoe races" (49). He'e nalu was historically, and remains today, a path of Lono, a surviving path through which present Kānaka Maoli can still obtain the ever vital mana in their lives. He'e nalu allows for both the physical and spiritual movement of Kānaka; it enables the most profound connection to and understanding of identity through place. This is a contemporary attainment of mana: being rooted while finding routes.

As a means of obtaining mana, tales of Kānaka Maoli engaging in he'e nalu, particularly ali'i such as Kūanu'uanu and Punahoa, infiltrate Hawaiian mo'olelo with images, tales, and songs about courtships, deathly wagers, and the spiritual joys of riding water. Some of the favorite surf spots of aliʻi have been recorded by Kamakau. He lists a few that the aliʻi favored on the island of Maui: "The chiefs of Wailuku passed their time in the surf of Kehu and Kaʻakau, those of Waiehu and Napoko in the surfs of Niukukahi and 'Aʻawa, while those of Waiheʻe were accustomed to amuse themselves in the surfs of Palaʻie and Kahahawai" (Kamakau 1992, 83).

Ka nalu in heʻe nalu also helped to uphold political hierarchies by establishing a kapu system within the sea, a system that designated who was and who was not entitled to surf specific breaks. Mary Kawena Pukui translates anʻōlelo noʻeau (saying) telling of a surf made kapu for ke akua Pele:

2356 O ʻAwili ka nalu, he nalu kapu
kai na ke akua.
*'Awili is the surf, a surf reserved
for the ceremonial bath of the
goddess.*

surfs at Kalapana, Puna: Kalehua, for
children and those just learning to surf;
Hoʻeu, for experienced surfers; and
ʻAwili, which none dared to ride.
When the surf of ʻAwili was rolling
dangerously high, all surfing and
canoeing ceased, for that was a sign
that the gods were riding.
(Pukui 1983, 257)

The kapu of oceanic regions did not just establish order in ancient Hawaiʻi; they signify genealogical place claims for contemporary Kānaka Maoli. Pacific Studies scholar Rob Wilson provides an interpretation of a Hawaiian name chant that links the ocean to protective ancestral gods, ʻaumākua, citing the ocean spaces occupied by these ʻaumākua as markers of Kanaka genealogical place claims. The name chant titled "Hula Mano no Ka-lani-opuʻu / Shark Hula for Ka-lani-opuʻu" dually praises high chief Kalaniʻōpuʻu and "his high-blooded genealogy as linked to the *ʻaumākua*, 'ancestral sharks' who served as protectors and as ties to the past" (Wilson 2000, 199).

Wilson describes the oli as articulating a "cultural-political strategy" in which the chief possesses land and sea rights through his links to a protective ancestral spirit, the "white-finned shark riding the crest of the wave" in the waters surrounding the land. More significantly for contemporary Kānaka Maoli is Wilson's interpretation of this oli as showing that Kānaka historically viewed the waters flowing around the edges of their islands as part of the same territorial space to which they are genealogically connected. From the many significations of land and ocean in Kanaka contexts emerge long-standing ties to indigenous Polynesian geographic rights as well as perceptions of spirit, heritage, and place (199). Yet both aliʻi and makaʻāinana actively participated in heʻe nalu, and both had access to sections of ke kai, which is significant because this access enabled the development of an oceanic literacy that reaffirmed an omnipresent ancestral identity in relation to ʻāina.

The politics, history, knowledge, and values preserved in these moʻolelo and oli are engaged every time a Kanaka surfer enters the time and space of ke kai. The surf breaking along the shores of Hawaiʻi resonates with those tales of love soaked into the sand, stories, for instance, about the love triangle between Māmala, Honokaʻupu and ʻŌhua at the surf called Kalehuawehe, known today as Waikīkī; the moʻolelo of Hauaʻiliki impressing Lāʻieikawai with his surfing abilities; and the tale of the beautiful chiefess and surfer Keleanuinohoanaʻapiʻapi from Wailuku on Maui. Kamakau recounts in the moʻolelo of Lono-i-ka-makahiki about how heʻe nalu often played a significant role in courting a young man or woman: "Kama-lala-walu, ruler of Maui, met him [Lono] and welcomed him royally. The chiefly host and guest spent much time in surfing, a sport that was enjoyed by all. It showed which man or which woman was skilled; not only that, but which man or woman was the best looking. It was a pleasing sight, and that was why chiefs and commoners enjoyed surfing" (Kamakau 1992, 53).

One, if not the greatest, love triangle in Kanaka history involves heʻe nalu. Turning to the epic moʻolelo of Hiʻiakaikapoliopele, we learn how Pele sends her little sister, Hiʻiakaikapoliopele, to fetch her lover, the aliʻi of Kauaʻi, Lohiʻau, to bring him back to her on the island of Hawaiʻi. When Hiʻiaka finds Lohiʻau, however, he is dead. Hiʻiaka brings him back to life, and the first thing Lohiʻau asks to do after his strength is regained, is to go surfing. Hiʻiaka creates a great storm, bringing waves from Kahiki for surf. The performance of heʻe nalu on the great waves at Hāʻena created the first bond between Lohiʻau and Hiʻiaka, a bond that grew and enraged a jealous Pele. Suspecting an affair, Pele destroyed Hiʻiaka's beloved ʻāina as punishment for her mistaken disloyalty, which in turn drove Hiʻiaka into Lohiʻau's arms, an affair that Lohiʻau paid for with his life.

Contemporary Kānaka may not know all of the specific moʻolelo telling of the political, economic, and social power churning within the sea. Hawaiian moʻolelo also vary from island and to island and region to region, and there are no distinct teaching methods for oceanic literacy in Hawaiʻi, with different schools of surfing styles and wave reading. Reading oceanic literacy today is a spatial engagement, a performed pattern that represents anew, a kinesthetically and visually documented event in the Hawaiian sea. Heʻe nalu writes this language into existence, transcribing history and culture through an ephemeral act (Carter 2009, 123).

Engaging in the enactment of heʻe nalu enables Kānaka to remember and (re)create vocabulary, concepts, images, and modes of reasoning with which to perceive and understand oneself through a cultural frame, and to then respond emotionally, seek intellectually, and creatively articulate one's individual views. Kanaka surfers don't just slide into states of remembrance in the sea; there is also a creative dipping, shifting, quivering, and cleansing. These sensations link the mind and body to a deep identity moored in a sense of belonging, a very powerful concept for human beings. Understanding oneself as related to the world focuses an acute spiritual awareness. Blankenfeld says, "You're just doing things so naturally, so effortlessly, so mindlessly that everything just flows [when surfing]. And what it comes down to is just pure joy. You do things, and you're already thinking five steps ahead. It's already a part of you, so it's pure joy. You watch hula dancers, the real good ones, it's an expression, of themselves, of their teacher, of their school of their culture. And they're just finding such magic in being able to express it so beautifully like they've been taught, and have taken pride" (Blankenfeld 2008).

From these connections, through this enactment, a specific understanding of identity is strengthened, and the mind, body, and soul align in a working rhythm of fluid movements that express a tranquil rush of potential. Individuals are capable of more when they draw from an infinite source of energy, literally balancing upon and within these forces, both harnessing and submitting to them. In heʻe nalu, a surfer must push herself over the edge of a breaking wave, hurl the body into the sound of white explosion, to ride its power. This commitment to an affective awareness and environmental consciousness of a multifarious existence within a fluid and constantly expanding place is the essence of seascape epistemology.

Rhythms in Seascape Epistemology

In the performance of heʻe nalu, a Kanaka surfer's center moves with the sliding ocean. Tapping into its energy, she relinquishes control by immersing herself in the mercurial place of the sea. Seascape epistemology is this organic dance between surfer and sea, this exchange of identity and bodily contact: warm skin on wet water. The surfer's "center," a pivot point, finds a rhythmic alignment within the wave; rhythm being the contraction of movement into physical forms, as Carter puts it. The surfer's center makes sense of both movement and stasis. She engages what Carter terms an "eido-kinetic intuition" as she is "stitched into the passages of the world" by entering the moving pathways in ke kai (Carter 2009, 15). The surfer's moving center becomes the stable point as she glides above and around dominant and seemingly immoveable lines drawn upon the world. The indigenous performance of heʻe nalu within a seascape epistemology brings the body back into places. Attention is tuned into the tempos of surging tides, migrating sharks, and circulating billows—all moving in an interconnected rhythm, telling the body how to travel through them. Seascape epistemology, as seen through the literacy of heʻe nalu, demands a center that is always moving, seeking to grasp a mobile but determinate complexity. The body is the central point of contact in an oceanic literacy; the body is the first point of analysis, the point of contact, the tool for ensuing investigations.

As the rhythms of the seascape assemble with the body, identity becomes an affected process over time and through philosophical, spiritual, and kinesthetic interactions with place. An ocean-body assemblage emerges. The term "ocean-body assemblage" is built from Manuel DeLanda's assemblage theory, in which an entity is defined by the totality to which it belongs and by its relations to other parts in the whole (society). This assemblage honors an indigenous relationship with the environment, rejecting "anthropocentric perspectives," and favoring fluid movements, or rhythms, which pull together the self with the world (DeLanda 2006).[2] According to DeLanda, of the three forms of matter—solid, liquid and gas—liquid has the most potential to create because it is constantly on the edge of chaos. Herein lies the potential for a surfer to self-organize a complex identity: Kanaka and American, artist and oceanographer, logical and emotional, all while moving through a liquid form of cultural genealogy, in distinct yet improvised rhythms.

As the body merges with the seascape into an ocean-body assemblage, ways of knowing and being are opened up to innumerable ways of moving, pausing, constructing, and deconstructing tempos as Hawaiian rhythms of cultural sovereignty are both disrupted and enabled. For instance, explosive bombs are still detonated by the U.S. military in Mākua Valley on Oʻahu, a sacred site of Kanaka Maoli *iwi* (bones) and *heiau* (sacred and worship sites). The army has been present in the valley since the 1920s, occupying thousands of acres after the Japanese attack on Pearl Harbor. A reterritorialization, however, is simultaneously occurring as Kānaka Maoli make visible and audible a Hawaiian resistance, such as through Anne Keala Kelly's film *Noho Hewa: The Wrongful Occupation of Hawaiʻi.*

Kelly illuminates the omnipresence of and profound impact that the U.S. military has in contemporary Hawaiʻi by giving facts: the military controls 20 percent of the land in the Hawaiian Islands; it has a population of over 11 percent in the islands, as opposed to less than 1 percent of the U.S. population as a whole; and the army has secretly tested chemical, biological, and deadly nerve gas agents in Hawaiʻi. Kelly argues that the military occupation of the Hawaiian kingdom in 1893 set in motion all of the colonial social, economic, and political structures predominant in the islands today that have led to a reality in which Kānaka Maoli are among the largest ethnic groups in terms of homelessness, imprisonment, health problems, and poverty levels. Kelly's film received national attention and won the Hawaiʻi International Film Festival's "Best Documentary" award in 2008.

Kelly represents one of many Kanaka artists and intellectuals continually (re)creating rhythms of Hawaiianness, offering a counterpolitics by (re)imagining a historical memory of being in spite of and alongside a colonial history of being (Shapiro 2000). Like Kelly's art, a historical memory of being reactivates when I dance in the sea at Mōkapu with Kūʻau. Reenacted is a memory of movement. Kānaka Maoli have been paddling *papa heʻe nalu* (surfboards) and waʻa beside Kūʻau for centuries. The braided waves have run through this little bay for hundreds of thousands of years, chipping away at its shores with an intimate battering. Kānaka have always engaged these waves, and their paths are now being retraced by me, as they will be retraced for generations to come. Heʻe nalu becomes an act of remembering a history otherwise forgotten outside the experience of the contemporary Kanaka surfer (Carter 2009). These memories, like the bodies and places that carry them, don't just "stoically endure"; they move and evolve with space and time (Lehrer 2007).

A contemporary Kanaka identity, remembered by me when surfing, as related to my cultural origin of the sea, has come to include the reality of those modern edifices that are in sight. Sitting in ke kai at Mōkapu, my vista now also includes a military tower, power lines, and trails of asphalt leading up the ridges. The scars from America's militarization and industrialization projects run deep, but this offshore perspective continues to provide Kānaka Maoli with a distinctive view from the views one experiences when standing on dry land. The signs of capitalism and militarization appear less dominant when sitting at the skirt of bounding cliffs rising out of the reef. Absent from this view are the colonial experiences within a society that have appropriated Hawaiian identity and culture. I am no longer subject to a tourist's gaze from this locale; there is no expectation of being a native surfer girl from the Hollywood film *Blue Crush*. The scorn of an English teacher for a student speaking pidgin (a Creole language that dominates the islands today, and which developed out of the need for communication between the different ethnic immigrant workers) cannot float in the ocean, nor can the opposing judgment from the Hawaiian community toward that student for being too Westernized. Privileging memory over history ensures movement and possibility despite the ways in which memory is shaped by the "now." Colonial realities are still in sight, but their busy gestures are diminished and blurred, drowned out by the bellowing sea. This is the political potential of the historical and cultural knowledge of oceanic literacy within seascape epistemology for contemporary Kānaka ʻŌiwi; an empowerment that stems from a native ontology rooted in a Hawaiian relationship to ʻāina.

Ontological Affectivity

Surfing beside Kūʻau, I construct my own conception of existence through my phenomenological insight and methods as connected to place, thus allowing for an imagined existence in the world based on my embodied involvements within it. My stomach and jaw tighten as the horizon balloons and the sucking sound of water scratches in my ears: anticipation, anxiety, elation, pride. Human affectivity, as the perception of bodily changes, as well as the emotional responses to these changes, are what give meaning to each transitory shape and gesture of the forming wave. Affectivity helps connect me to the conglomeration of water molecules spilling toward me, which (re)creates new relationships and ways of existing in the world.

As Walt Whitman observed, the body and soul are inextricably "interwetted." Interpreting Whitman, Lehrer notes that emotions are generated from the physical body rather than the mind (Lehrer 2007). More significantly, Whitman concluded that these emotions influence and are required for logical thought. He wrote in "Starting from Paumanok":

Was somebody asking to see the soul?
See, your own shape and countenance, persons, substances,
beasts, the trees, the running rivers, the rocks and sands.

All hold spiritual joys and afterwards loosen them;
How can the real body ever die and be buried?
. . .
Behold, the body includes and is the meaning, the main
Concern, and includes and is the soul.
(Whitman 1992, 17)

Neuroscience now supports Whitman's observations, and has determined that our emotions, which are rooted in our muscles and organic palpitations, are an essential element of the rational thinking process. The emotion of fear, for instance, stimulated by the sea, ensures respect while also encouraging an acute and constant awareness of the ocean's power and potential. Fear is what allows for joy and the ability to slip across waves in a reciprocal relationship of exchange and interaction. The dual emotions of fear and joy create new experiences and thus a new potentiality.

Whitman articulated an ontological emotion in his poetry that illuminates how, in fact, the brain is embodied (Whitman 1992, 17). The corporeal aspect of everyday life is illuminated: an affected body creates subjective emotions that deconstruct relationships and knowledge, allowing for a reconstruction of relationships and knowledge. A Kanaka Maoli body is affected by its whole history of relations with ka moana, just as the sea is affected by its whole history of relations with the body (and the whole history of human beings), including human emotions.

For example, a navigator at sea is constantly processing information as he reads ka moana throughout the day, weeks, and even months, making determinations based on his body's memory of that information. He must make decisions based on the embodied knowledge accumulated in his nose, on his skin, and in his naʻau, or gut. His conscious and rational mind leans on cellular knowledge, intuition, and a kinesthetic reading of and reaction to the signs that are retained as embodied engagements. Oceanic literacy is invested within both the conscious and the unconscious, within both cerebral and embodied sources of knowledge, which inform the navigator how to (re)act.

A disembodied ontology fails to capture the collective experience of identity because the body would be left out of an understanding of self. Yet this is precisely how Euro-American reason, and how European theory, functions—by privileging the cognition of the mind over the movement of the body. Carter observes, "Like photographers taking care their shadow does not get into the picture, we absent ourselves from the scene of discovery. A description of the world is accounted most authoritative when it contains no trace of the knower" (Carter 2009, 5). Omitting the body creates hierarchies and forms of power that come to constitute realities that, intentionally or unintentionally, marginalize indigenous relationships to place and impoverish indigenous identities. Neglected are the memories held within kinesthetic movements, which, Carter explains, occur "in-between the makers of marks and the marks they make" (4), and which erase all the journeys that bodies have taken in the seascape: the knowledges accumulated, stories told, and discoveries made.

Ontological affectivity, however, enables the simultaneous movement of mind and body, of passion and logic, of idea and gesture. Ontological affectivity becomes a way of shaping our way of being-in-time and in-space so that our habitation in the world includes a responsibility to place and a profound understanding of identity within this relationship. Embodied practices such as heʻe nalu that directly engage the ʻāina through an epistemology that reaffirms indigenous ontology allow Kānaka Maoli to draw ancient as well as innovative, and communal as well as personal, connections to culture, history, genealogy, spirituality, and general happiness. The ocean-body assemblage that comes out of an ontological emotion (the physiological engagement with the sea as well as the philosophical understanding gained through the phenomenological insight of oceanic literacy) becomes an empowering framework for contemporary Kānaka Maoli that is privileged, even required, by seascape epistemology.

Ontological Time

This dynamic flow of material, bodies, and memories in the seascape represents the evolution (deconstruction and reconstruction) of knowledge and identity through time. For Kānaka Maoli, time is created through place in such a way that place, in fact, creates time, and neither place nor time are void of Hawaiian bodies because the sand crabs and taro plants are genealogical manifestations of self, which are also place, and which create time. This ontological spatio-temporal construction is an embodied memory that does not seek to represent places or events in abstract time, but instead to engage time through and as a construction of place in relationship to a construction of identity.

The significant aspect of this spatiotemporal construction is that it renders "identity" as never complete but rather an ongoing engagement and process. Paul Ricoeur's concept of a "narrative identity" allows one to "think through the question of 'personal identity' in a new way, taking into account the temporal dimension (the temporality) of a being who, by existing with others in the horizon of a common world, is led to transform himself in the course of a life history, that is, who is what he or she is only in the course of becoming himself or herself," explains scholar Maria Villela-Petit (Villela-Petit 2010). Identity is always moving in relation to one's "time narrative," in relation to one's deep rhythmic thoughts. Ricoeur explains that an individual understood as a "who" of a "history," as part of a larger connection and the one upon whom the story confers a sort of identity, is an individual whose temporization shapes itself in accordance with a narrative model. Part of ontology is geopolitical time.

Take, for instance, the time narrative of me, a Kanaka surfer, en route to and through the sea. I must exist in the time of the state as I drive to the ocean, down the street, in a car, stopping at red traffic lights and adhering to painted crosswalks. I move through these imagined territorial state boundaries that require legal recognition: driving on the right side of the road, adhering to the speed limit, following street signs. I, however, might pull off the road and walk into a Starbucks Coffee shop, leaving the time and space of the state in which I am a citizen to enter a capital time, a time of consumption in which my consumer identity emerges. A capital time is actually a countertemporality to the time of the state because, as Michael J. Shapiro notes, it is a time that cannot be contained by the state (Shapiro 2000). Inside capital time, space becomes increasingly linear and focused as it distinguishes between the past, present, and future, designating the movement of time as unrelated to anything but the given moment (Alliez 1996).

Waiting in line at Starbucks, I might answer an incoming call on my cellular phone, engaging me in two simultaneous times: a time of capitalism and a time of communication technology, which is also a countertemporality to national time. Sipping on my coffee, I re-enter national time as I put quarters into a state-erected parking meter across from the beach. But as my body enters the sea, I enter an indigenous thought-world stimulated by cultural memory, imagination, perception, and understanding. Time becomes a condition of action in the ocean. Privileged here is a Kanaka time that washes away the linear times encountered along my way to ke kai. This shift into Kanaka time offers me a historical depth to how I imagine myself. Entering the time of the sea alters the locus of power, honoring a Hawaiian ontology and epistemology as opposed to my state-imposed identity as American citizen.

My identity as American citizen is understood through a shared historical process between myself and the United States: missionary arrival, plantation time, annexation, statehood. These processes are also temporally disjunctive; they simultaneously conflict and coexist. Kanaka time is juxtaposed with the historical time of the State of Hawai'i to show the diverse forms of copresence that are, as Shapiro says, "derived in the discourses of nation-state legitimization." Shapiro notes that ontology and epistemology always engage in political interactions as a means of continually negotiating the potential copresence between those with diverse ways of being-in-time: American citizen and Kanaka Maoli (Shapiro 2000, 79). Because time and space cocreate a distance, which can carry over to a theoretical absence by keeping the Other's time outside of the system, temporal and spatial imaginaries are necessary within a modern geopolitical reality so that Kanaka can expand ways of being-in-the-world and being-in-time to include our own.

Understanding one's autonomous identity, even as the body moves in and out of different times, requires an understanding of the sense in which one is oneself in time, because individuals exist in different time narratives, at times combining identities simultaneously, depending upon political, economic, cultural, and environmental factors. Ricoeur's time narrative enables this multisitedness, this representation of different experiences of time outside of purely linear or chronological conceptions, because, as Atkins explains, "A narrative may begin with a culminating event, or it may devote long passages to events depicted as occurring within relatively short periods of time. Dates and times can be disconnected from their denotative function; grammatical tenses can be changed, and changes in the tempo and duration of scenes create a temporality that is 'lived' in the story that does not coincide with either the time of the world in which the story is read, nor the time that the unfolding events are said to depict" (Atkins 2003).

Including Hawaiian time in my time narrative, I allow my indigenous epistemology and ontology to emerge. I enable an oral and embodied narrative to be told in cyclical rhythms in which the past, present, and future are simultaneously employed, often becoming interchangeable. 'Ōlelo Hawai'i expresses the past as *ka wā ma mua* (the time in front or before), and the future as *ka wā ma hope* (the time after or behind). From the Kanaka perspective, importance is held in what came before, in the past rather than in the future, because the past is a time of knowledge, while the future remains unknown. It is the past that shapes the present and future, which is why genealogy is so predominant in Hawaiian culture.

Kame'eleihiwa explains, "The genealogies are the Hawaiian concept of time, and they order the space around us" (Kame'eleihiwa 1992, 19). Kanaka ancestors, traced back to the gods, gave birth to the world and to the beginning of time itself. Kanaka time begins with Pō, female night, and she is the ultimate ancestor to Kānaka Maoli and to all the Hawaiian gods ("from the night" means "from the beginning of time," *mai ka pō mai*). Pō spontaneously gave birth to Kumulipo, a son, and Pō 'ele, a daughter, and these two siblings mated to create the gods and the world. Their first child was the coral polyp, an akua named Hina, whose body was the space consisting of the coral reefs. Hina in turn gave birth to the other sea creatures, such as the coral insect, earthworm, and starfish, as well as Kū'āu, the *pōhaku* (stone) in the sea.

Within this narrative, Kanaka time is perceived through an ontological understanding of knowledge and identity in relation to ka 'āina, which circulates in space. Kānaka Maoli, Kanaka ancestors, and ka 'āina are all connected through an embodied memory that can be recalled through seascape epistemology. Ka 'āina, the land and the sea, creates an interarticulation between the self, aesthetics, and politics for Kānaka Maoli. The seascape becomes political as it speaks to the aesthetic rhythms and idioms of Kanaka dwelling, and of the Kanaka embodied connection to the sea. These alternative spaces and places created between Kānaka Maoli and the seascape not only presume an indigenous epistemology embedded in an indigenous ontology but also alter the loci of power, speak to colonial resistance, and offer an experiential place from which to frame ideas and conclusions about the world.

Jacques Rancière helps to explain how seascape epistemology activates this political construction of time. He explains:

Plato states that artisans cannot be put in charge of the shared or common elements of the community because they do *not have the time* to devote themselves to anything other than their work. They cannot be *somewhere else* because *work will not wait*. The distribution of the sensible reveals who can have a share in what is common to the community based on what they do and on the time and space in which this activity is performed. . . . There is thus an "aesthetics" at the core of politics . . . It is a delimitation of [14] spaces and times, of the visible and the invisible, of speech and noise, that simultaneously determines the place and the stakes of politics as a form of experience. Politics revolves around what is seen and what can be said about it, around who has the ability to see and the talent to speak, around the properties of spaces and the possibilities of time. (Rancière 2004: 12–13)

The political is about entitling people to speak about politics as opposed to being mere subjects of it. This is why politics is aesthetic for Rancière; it is created by those who do not have the time to engage in these "political" activities. He posits that the distribution and redistribution of times and spaces, places and identities, the way of framing and reframing the visible and the invisible, of telling speech from noise, is the partition of the sensible. Thus, politics consist of the reconfiguration of the partition of the sensible, in bringing onto the stage new objects and subjects. A politics of time is illuminated, as some have a "lack of time" to speak and participate, and are thus left "outside" to become subjects of politics (Rancière 2004).

In this way, seascape epistemology engages a politics of aesthetics. Despite marginalization, seascape epistemology (re)engages Kānaka Maoli in relationship with and knowledges of the seascape, a time and space that holds a Kanaka sense of being, history, and economy. Modern Kānaka haven't had as much "time" to engage the ocean as they did prior to colonization because they struggle to pay rent in a "paradise" housing market. Seascape epistemology is a politics of aesthetics in that it redistributes times, spaces, places, and identities from a focus on colonial power, disruption, and displacement to one on the reconnection and reconstruction of an autonomous identity.

The time narrative of Hawaiian genealogy becomes a political tool, giving contemporary Kānaka location through their relationships to ancestors despite emphasis placed on colonial realities that dominate the time of "today." Kānaka stand in the space of ancestors who constructed the significance of these spaces through time. The Hawaiian time narrative enables Kānaka to construct autonomous spaces in the world as opposed to being captives or in someone else's construction of space. Native Hawaiians draw from genealogy to create meaning and connection in the world, phenomenologically constructing a self-determined epistemology and ontology within a reality that has marginalized Kanaka spaces and times. As Kānaka Maoli know who and where they are in the world, through a genealogical connection to ancestors and the geographic space of the Hawaiian Islands, time is bound to Kanaka identity. Hawaiian time is Hawaiian genealogy, but time is also a specific relationship to this genealogy of grandparents, gods, and the ʻāina. Kanaka time is the time of stories as both individual and collective narratives in the past, present, and future, and thus a Hawaiian identity is never stagnant but always (re)creating.

Time is not merely a cyclical narrative, however; it is also the rhythm of this cycle. For Kānaka, the cyclical rhythms of the days, nights, tides, winds, marine animal migrations, seasonal plant flowerings, and rain patterns function as ways of discerning time. In early Hawaiʻi, fishing and ocean gathering were carried out according to the moon phases and the stars. During the *huli* (change) from *hoʻoilo* (the wet season) to *kau* (the dry season), *limu kohu* (a native seaweed) dies on the reef from sun exposure, telling Kānaka Maoli this is the time to pick as much limu as desired. When the stars were numerous and bright, that was the time to go and look for the shellfish such as *kūpeʻe* (*Nerita polita*), which usually hide during the day (McGregor 2007, 93). Summer was the spawning time for the *manini* (reef surgeonfish), *humuhumunukunukuāpuaʻa* (trigger fish), *ʻamaʻama* (mullet), and *āholehole* (flagtail) fishes. These would be caught and salted to provide food throughout the year. The baby *ʻoʻopu* (goby fish), called *hinana* would hatch and develop in the salt water from August or September through November (109).[3]

The time of the sea employs the practical and critical knowledge of fish cycles, safe passage into and out of bays, and knowledge of star patterns for navigating. It is a spatial temporality that when employed as an oceanic literacy offers not only psychological decolonization but true physical decolonization. If Kānaka Maoli know how to exist in the world independently of and alternatively from the dominant system, Kānaka can survive outside of it. This becomes particularly critical for homeless Kānaka ʻŌiwi living on the beach, who have been marginalized by the dominant time of capitalism, a system in which the U.S. dollar is a value being valued in itself.

Money capitalizes the essential dissemblance of potential time because, as Éric Alliez explains, it reduces space and time by isolating value into a reduction of production and reproduction, with money being an instrument of production. Capital time becomes a moment of clashing temporality for homeless Kānaka Maoli.

Recognizing that the issue of homelessness is more complex than I address here, it is still important for me to mention that a strong connection to ʻāina offers the homeless a powerful tool and an alternative time in which to survive by employing ocean-based knowledges. Being-in-a-time of ke kai, an ecological time of collecting food, bathing, and finding shelter, engages a more complete study of time as a function of its history unfolded within thought-worlds, and according to times that are not necessarily congruent. Recognizing indigenous time alongside dominant temporal systems such as capital time enables a plurality of identity, and in the case of Kanaka homeless, an alternative political economy. Time is remobilized through oceanic literacy; Kānaka are ontologically affected and aware, resisting the identity-fixing effect of a state-oriented model of political space as well as the homogenization of the temporal presence of citizenship of linear time (Shapiro 2000). In ka moana, Kānaka Maoli are engaged in a time that allows an indigenous identity to become visible and viable within dominant discourses of identity and time.

Ontological Space

As seascape epistemology requires multiple and seemingly disparate times in which to exist, it also creates diverse and indigenous spaces in which to move. Imagine the Hawaiian sea moss, *limu pālahalaha*, crawling between the cracks of a concrete jetty. The moss is able to maneuver its way between the crevices of the concrete, creating alternative pathways atop and within an imposed structure (colonization). The limu pālahalaha exists in a "smooth" space, which, defined by Gilles Deleuze and Felix Guattari, is nonmetric and acentral, a directional path that can rise up at any point and move to any other point, like a formless and amorphous space (Deleuze and Guattari 1987). The concrete jetty, however, exists within a "striated" space that is limited by the order of its own space or by preset paths between fixed and identifiable points. The smooth space of the moss, and of the seascape itself, allows for diversity and for the emergence of marginalized identities simply because it allows for movement between the lines.

A politics of aesthetics emerges as the limu pālahalaha not only creates but also makes visible new spaces within the striated space of the jetty. Time and space configurations are ideological constructions of power within epistemology, and because they control constructions of identity in the global arena, or state-established and recognized ways of being-in-time and space, the smooth space of the sea offers itself as a means of disrupting power structures. It is not the momentary location in identifiable space but the continuous movement in amorphous space that is critical to understanding the political opportunity of seascape epistemology as well as human existence in time. The space of the ocean is an "occasion," a moment that comes about through the experience of sensation. The fluid enactments of oceanic literacy are what allow Kānaka Maoli to be part of the smooth space, to create an autonomous identity by engaging a discourse on colonization through movement.

Striated and smooth spaces, however, cannot exist in isolation. They must always coexist and intermingle. Ke kai is a smooth space of autonomous creation for Kānaka Maoli, because as a space that holds specific social meaning, the ocean is simultaneously also a striated space for Native Hawaiians. The sea can become striated as it is coded and detailed to those familiar with wave shapes, speeds, and frequencies. The sea can be divided by Kānaka into specific fishing locales that might have seasonal *kapu* (prohibitions) during spawning times, for example. Even the smooth space of the sea allows itself to become striated. Both smooth and striated spaces can be used autonomously by Native Hawaiians to reaffirm identity and indigenous knowledge. Kānaka reinterpret the striated spaces of fishing holes as codes through Hawaiian contexts, and they can summon a specifically Hawaiian world not subject to a dominant system of reference.

Many oceanic striations visible to Kānaka Maoli are elusive to those unfamiliar with ke kai. The tourist may observe a surf break as monochromatic lines of waves rolling into shore, while a Kanaka surfer takes note of spatial striation within this break: where the break is divided into entry and exit points that depend upon the strength and direction of the current that day, as well as the direction of the swell and tide. The waves breaking close to shore are understood in terms of form and function, of how to get through and out and then back. Waves breaking outside indicate where the surfer needs to position herself to ride.

There are also temporal striations understood by the Kanaka surfer. Immediately entering the ocean, I know my skin will be physically refreshed from the heat of the sun, but after several hours of saturation, it will wrinkle with a chilled bend. When I first dive into the sea, my feet kick like sinking logs, but soon begin to remember their aquatic abilities, webbing with the efficiency of fans sweeping through water. Over time, I more easily attune myself with oceanic movement. I am more receptive to and aware of slight changes in the tide: a piece of reef I align myself with all day moves closer to my dangling feet. I notice the modest change in the wave's shape, now steeper and faster as it breaks just above the coral.

Oceanic striations are also created by state-designated boundaries. I paddle out past a delegated swimming-only zone in the ocean on my way to an appointed surfing zone, both artificial divisions established by the Hawai'i Department of Land and Natural Resources. I might have to navigate zones seized by capitalism, such as surf lessons, usually areas closer to shore where waves are smaller and more forgiving. I might encounter a zone "fenced off" for exclusive use by a surf contest (contest promoters purchased these striated times and spaces from the state and county). Grids have been formed that identify space within the ocean as being specific to one purpose, as being this and not that, designating belonging and exclusion.

Artificial grids and boundary lines, however, can be "deterritorialized." Deleuze and Guattari would state that the limu pālahalaha that grows on the concrete jetty, while dictated by the shape of the jetty, "deterritorializes" the concrete (Deleuze and Guattari 1987). The moss alters the space of the jetty from a geographic zone to a feeding ground for fishes and crabs. The meaning of the space initially created by the concrete jetty is altered by the limu. Striated spaces become smooth, just as smooth become striated. The ocean, inherently from within its own body, enables plurality.

This process of political and social deterritorialization is particularly pronounced in Palauan fishing lagoons. R. E. Johannes writes, "Traditionally a chief is expected to be a good fisherman, but possesses no special authority and receives no special treatment while fishing. When Palauan fish, land-based protocol is suspended. Harsh criticism, or 'words of the lagoon,' tekoi i'chei may be hurled by man or boy of any rank at anyone, chief included, whose efforts do not measure up on the fishing grounds. . . . Thus has excellence in fishing been reinforced for centuries" (Johannes 1981, 3). In the specific smooth space of the lagoon, in the time of fishing, the world is different. It is a different experience of time and space, of interaction and power relations. The nucleus shifts, and the political and social boundaries are deterritorialized.

The space of the sea is not political in itself, however. As Deleuze would state, the virtual potential of the sea becomes actualized through seascape epistemology. Spaces of possibility, in which knowledge and identity sit, come about only through the constructed meanings within these spaces. Kanaka ontology is not about a spatial location in the Hawaiian Island archipelago, but about the ways in which Kānaka ʻŌiwi dwell within this specific spatial location. Native Hawaiians are located within a space of possibility, which provides social meanings. Martin Heidegger posits that how we dwell speaks to how we know, enabling specific places to create a larger network of associations with political and ethical implications.

According to Heidegger, the condition of human beings and the world is bounded in time, and this temporality is grounded in an origin. Heidegger asserted that one cannot "be" in an abstract sense without being in and of a particular place, situation, or context: being-in-the-world, or being-there, *Dasein*. Building upon this philosophical perspective follows the concept of seascape epistemology. Seascape epistemology is a temporal epistemology embedded in a metaphysical ontology. Because the seascape holds Kanaka genealogy, history, spirituality, and economy, and is thus a reference point into our past (and future), Native Hawaiians are embedded in a metaphysical nature of being that is intertwined with the sea. Hawaiians and the seascape cocreate an emerging series of places as the seascape is incorporated into Kanaka identity and Kanaka identity is incorporated into the seascape. The seascape becomes political as it speaks to the aesthetic rhythms and idioms of Kanaka dwelling, and of the Kanaka embodied connection to the sea.

Heidegger perceives that what it means for people and things to "be" has little to do with the fact that we are "in the world"; it has little to do with the fact that we and the world are inseparable. The idea of the world as something to be known, that there are categories of knowledge, or the question of what it is for something to be a thing is not the issue for him either. Heidegger stresses the fact that we are not knowers or spectators in an isolated and independent sense; instead we are "engaged" in the world. This is crucial to the question about how epistemology is embedded in ontology.

What we're involved in is not knowledge in the abstract sense, Heidegger states. The world is not "things" or "things we know" but is spoken of in terms of the "equipment" with which we engage it. Being-in-the-world for Heidegger is about being engaged in tasks with "knowledge," and focusing on the task at hand (Dreyfus 1991). Being engaged in tasks involves an unconscious type of reflection. It is important to note that Heidegger wants to minimize the role of the conscious subject in his analysis of human "being" by asserting that the basic relation between the mind and the world is a relation of a subject to objects by way of mental meanings. He wants to minimize the role of the conscious, because if one is really engaged in work—those "structures" of surfing over a shallow reef, with a board carved from a koa tree, by reading the swell direction and incoming tide, with the purpose of courting a young aliʻi who has caught your eye, or those "structures" of fishing off a reef, with a net, by observing the ripples and colors moving across the water's surface for hours, with the purpose of helping to feed an ʻohana, or extended family—should be both transparent and hidden in background practices and skills. If we are truly engaged in the action, the equipment won't be analyzed, it will simply function as we need it and unconsciously know how to use it.

The notions of territory and spatial imagination become part of ontology. Kanaka approaches to knowing are related to place and are ocean-based because Kānaka Maoli were historically surrounded by the sea—and most still are. Kānaka speak of identity as coming from the land and sea: "child of the frothy sea," "the aliʻi of Kona," or "the people of Mākua." Everything begins with the ʻāina: this valley in Hāʻena, Kauaʻi, is where Hiʻiaka fetched her elder sister, Pele's lover, Lohiʻau; this bay, ʻUo, is where the famous waves were ridden by aliʻi Kūanuʻuanu; this rock at the point of Kaʻena is the manifestation of the evil deity Pōhakuloa. ʻŌlelo Hawaiʻi and moʻolelo speak from the land and sea first; everything starts from the ʻāina, all thoughts and notions of existence, history and knowledge, come out of place.

Keith Basso, a culutal and linguistic anthropologist, investigated how the Western Apache imagine space, and how that imagination constructs their ways of being-in-the-world. He asked, "What do people make of places?" realizing that human beings have attachments to them as a connection between their identities, sensibilities, and origins. Basso wrote,

> Through a vigorous conflation of attentive subject and geographical object, places come to generate their own fields of meaning. So, too, they give rise to their own aesthetic immediacies, their shifting moods and relevancies, their character and spirit. Even in total stillness, places may seem to speak. But as Sartre makes clear, such voices as places possess should not be mistaken for their own. Animated by the thoughts and feelings of persons who attend to them, places express

only what their animators enable them to say. . . . Human constructions par excellence, places consist in what gets made of them—in anything and everything they are taken to be—and their disembodied voices, immanent though inaudible, are merely those of people speaking silently to themselves. (Basso 1996, 108–9)

For the Apache, specific places are identified with their ancestors or sacred histories, and these sites act as guides to how to behave or exist. Thus, place is firmly embedded in moral imagination for Apache, in their ontology and ethics. Dudley, Basso's Apache informant, told him, "Wisdom sits in places. It's like water that never dries up. You need to drink water to stay alive, don't you? Well you also need to drink from places" (127).

Edward L. H. Kanahele, later professor at Hawai'i Community College and cofounder of *Hui Malama I Na Kupuna 'O Hawai'i Nei* (Group Caring For the Ancestors of Hawai'i), an organization which aims to protect Kanaka burial practices, also relates his relationship with imagined space to his identity, sensibility, origin, and a set of ethics. He explains how the understanding of place, place names, and the designation of places as sacred, *wahi pana* (legendary place), are critical to understanding the relationship between place and Kānaka Maoli:

> As a Native Hawaiian, a place tells me who I am and who my extended family is. A place gives me my history, the history of my clan, and the history of my people. I am able to look at a place and tie in human events that affect me and my loved ones. A place gives me a feeling of stability and of belonging to my family, those living and dead. A place gives me a sense of well-being and of acceptance of all who have experienced that place.
>
> The concept of wahi pana merges the importance of place with that of the spiritual. My culture accepts the spiritual as a dominant factor in life; this value links me to my past to my future, and is physically located at my wahi pana. (Van 1991, ix)

Kanaka indigeneity is rooted in and routed through this specific place, ka 'āina. Situated inside places, Kanaka bodies can move like the limu pālahalaha, continuously morphing to adapt to the modernizing (and diasporic) seascape, autonomously creating Kanaka times and spaces in which to exist. An ocean-body assemblage enables identities to swirl in water and dance with the skirt of a mountain. Knowledge and identity, when merged with the sea, is freed up and becomes a multiplicity of experience in a seascape epistemology.

Waves of Knowing

Knowledge is the subjective accumulation of what a person has absorbed, experienced, and imagined through sources of sound and touch, processes of sight, essences of taste, constructions of memory, and structures of language that create an emergent self able to articulate and expand upon what is already "known," how we know, and who is a "knower." Seascape epistemology exists within this realm as a way of knowing that honors the embodied processes within written, oral, painted, experienced, and historical literacies expressing diverse knowledges. Seascape epistemology is continually de-creating and re-creating within an ever fluctuating time, space, and place; it enables identity to be open to the plurality of thinking, skills, wisdoms, awarenesses, interactions, and emotions—all united into an informal yet progressive existence. Identity becomes mobile, which allows for "knowledge": the movement, articulation, and focus of all the senses to foster an inherent connection to and awareness of the ways in which the environment and the world moves. Oceanic-literate Kānaka Maoli have learned to pay attention to their senses and to the environment as sources of knowledge, to read them not just as information but as intelligences carried on the backs of waves, tales of fish and the rays of the moon, the "memory of all the approaches that constitute its movement history" (Carter 2009, 268).

The kinesthetic body is included in seascape epistemology through performance, while place takes on meaning as related to identity, and care is taken to recognize the inexhaustible shapes, gestures, and tongues within distinct waves of knowing. Although seascape epistemology resists imposing itself as a dominant epistemology, filled with static boundaries and systems that erase other times and spaces, it is also created from a specific place and with a specific intention. There is a poetics of experience in seascape epistemology, an embodied connection of "being-in" the sea and the world that is an aesthetic interaction, a process and production of forms with political and ethical implications. Seascape epistemology is predicated on action, articulating how one is involved in a world-making process that includes physical acts of displacement as well as metaphorical processes of replacement (13).

As a poetics of experience, seascape epistemology must be careful not to essentialize a Kanaka (or any) way of being-in-the-world. While seascape epistemology is anchored in genealogy, it also emphasizes instability, de-centering, openness, and antiessentialism (often dubbed poststructuralist, postmodern or postfoundational). Seascape epistemology, while rooted in time, place, and ontology, forever provides itself as an alternative to the dominant landscape because knowledge and identity is constructed anew each time the body is affected by the organic and unpredictable sea. The ocean will always provide an alternative smooth space, an unchartered course one can forge through power structures and divides. Seascape episte-mology is one way in which Kānaka Maoli can fulfill the potential within ke kai as a means of reinventing a people by making oceanic literacy not merely practices but critical ways of knowing and interacting with the world.

HOʻOKELE   Seascape Epistemology as an Embodied Voyage

Seeing, Smelling, Hearing Ke Kai

For Kanaka navigator Bruce Blankenfeld, the oceanic literacy of hoʻokele is a portal to access his Kanaka ontology and epistemology through a specific way of reading and interacting with ke kai. Standing on the platform of the waʻa, which sits in the vast ocean, on the planet, which floats in an expansive galaxy suspended in an immense universe composed of tens of thousands of other galaxies, Blankenfeld engages an ocean-body assemblage by expanding his identity to connect with his surroundings. "Those stars have to be a part of you," he explains (Blankenfeld 2008). Being situated within the seascape allows Blankenfeld to connect affectively through his spatial imagination by feeling and seeing the link between these earthly and heavenly bodies through his own body, a necessary link to read how the skies and universe move, so that he knows how he can and should move through them. Blankenfeld is encompassing a seascape epistemology through the performance of his oceanic literacy of hoʻokele; he is engaging in an image-making power through his spatial knowledge of the seascape, from the drumming that the ocean beats upon his skin, producing a specific grammar and cre-

FIGURE 4.1. Navigator Bruce Blankenfeld on the *Hōkūle'a*, c. 2007.
Photograph by Russell J. Amimoto.

ating a new axiom about motion predicated on kinesthetic and affective interactions.

When navigating, Blankenfeld becomes an aesthetic subject whose movement articulates Kanaka Maoli ontology and epistemology through the stimulation of his somatic senses: sight, smell, taste, sound, and touch. His body and the seascape interact in a complex discourse in which Blankenfeld is able to "see" his location in the world by reading the yellow stripes painted across the sky, hearing the direction of swells thumping the hulls of the canoe, tasting the water's salinity, smelling the cool north winds, and feeling the intensity of the sun's heat. "Seeing" is a physiological process that involves more than the simple recording of light and images. It is a selective and creative process in which environmental stimuli are organized into flowing structures that provide meaningful signs (Tuan 1977, 10). An accomplished navigator from the Society Islands, for instance, Tewake,

could "see" his direction from inside his hut on his canoe's outrigger platform by lying down to better analyze the roll and pitch of the canoe as it
marched in a distinct rhythm over the waves (Lewis 1994). Feeling how the
canoe moved and where the waves hit the hulls enabled Tewake to "see" his
direction.

Blankenfeld explains that when voyaging, "most of the seeing is inside.
So before you even leave [on a voyage], you have a very clear vision of what
the target is" (Blankenfeld 2008). Blankenfeld "sees" his destination as he
voyages toward it, even when he's weeks away from arrival, by responding
to sensational messages collected through his feet, in his nose, and his eyes.
Sight becomes corporeal: the feet, nose, and eyes all have a specific way of
seeing, and become emotions burned into muscles that are interpreted by
the mind. It is this constant interaction between the body and the mind
that generates a specific epistemological and ontological context, expanding
Blankenfeld's world to include an intimate interaction with the white foam
of open ocean swells that his canoe surfs, a relationship whipped in joy and
a freedom of weightlessness through cellular memory.

This duality between an analytical awareness and sensational imagination is what allows Kānaka Maoli to see Hawaiian spaces and times in
the sea for movement and empowerment. The ways in which Blankenfeld
physically involves himself with the ocean affects how he "sees" it. Seeing
becomes a political process: as Blankenfeld's sight is expanded through his
oceanic literacy, and so is his ability to think outside of a static mind-set, a
mind-set that tends to enforce one specific way of knowing, or seeing, and
that relies upon one specific reality.

Oceanic literacy engages a subjectivization, a transformation of the aesthetic coordinates, as Jacques Rancière would put it, or the ability to control
what is visible (or invisible) in a community by making necessary the assumption that we are all equal (Rancière 2004). Oceanic literacy becomes a
political and ethical act of taking back Kanaka history and identity through
a rhythmic interaction with place: the swing of tides shuffling sand, the
sharp tune of swells stacking upon each other at coastal points, the smooth
sweep of clouds pulled down by the wind. Rhythms don't just represent
the ocean; they constitute it as figurative layers. Merging the body with this
rhythmic sea enables a reading of the seascape's complex habits, as well as
all the memories created and knowledges learned within this oceanic time
and space but have been effaced by rigid colonial constructions of identity
and place.

Affective rhythms of sight are political visions for Blankenfeld because what he sees when voyaging is a structure of his imagination: light hitting the retina of the eye is immediately processed through the subjective brain. Author Jonah Lehrer notes that without the brain's interpretation of light flashing off the retina's photoreceptors, "our world would be formless" (Lehrer 2007, 105). Lehrer adds, "But this ambiguity is an essential part of the seeing process, as it leaves space for our subjective interpretations. Our brain is designed so that reality cannot resolve itself," requiring our imaginations to fabricate a context for this "reality" (107). Blankenfeld is harnessing and mobilizing this image-making power inherent in all of our brains by reaching beyond light reflection into a realm of seeing through intuition with the aid of environmental cues and personal experience. Blankenfeld dilates his sight by constructing a vivid yet malleable pathway toward a destination through his oceanic literacy: he can find his direction and location by extending his sight into his imagination.

Sight is also expanded through the oceanic literacy of lawaiʻa. A fisher, for example, is able to "see" through the aquatic distortions of light waves bending, refracting, and reflecting off fish scales in the water because she knows that she is entering a distinct material world. Her rational mind tells her that light travels in a straight line because this is her experience on land and in the atmosphere. The fisher knows, however, to think like the ocean and to bend the light refracting off the fish's shimmering body. In a linear mindset, the fish is seven feet away when measured in a straight angle from the fisher's eye to the fish. She would miss her target if she approached the sea with this logic. Thinking like the sea, in bends and dips, the fisher knows to aim perhaps two feet closer with her net, instinctively taking the refraction angle of the light into consideration. She is expanding her brain capacity by intimately interacting with an evolving and constantly modified place.

In Kanaka Maoli's Oʻahu version of the moʻolelo of Nāmakalele, or "Flying Eyes," Keanahaki, wife of Keawe, fishes by throwing her eyes into the ocean (Beckwith 1970, 199). The eyes are perceived as taking on a life of their own within the ocean, seeing things they might not otherwise. This literal throwing of the body into the sea was a very successful tactic for Keanahaki, for she caught many fish. Kānaka Maoli attuned to oceanic literacy understand that to look with the body's eyes is more than a physiological act, it is a metaphorical way of thinking and imagining in relation to a specific understanding of the world and with a specific relationship to it. The body and mind are potentially politicized as sight expands into a realm of a geopolitical cartography. The body and mind enter a realm that is geopolitical because it contrasts a Euro-American mapping of landscapes (militaristic, capitalistic, and touristic). A Euro-American mapping positions lines and dots across places that deny the fluidity of spaces, such as the seascape, to re-imagine mind, body, and place. A Euro-American mapping of, for instance, a shoreline as a sovereign and absolute phenomenon governable through realism, denies space along this shoreline in which Keanahaki's eyes can be imagined or understood.

A Kanaka fisher's sight is related to his ontological understanding of the world, which rests in a genealogical relationship to the sea and its creatures. Kānaka Maoli enter the hunt with a feeling of respect and *kuleana*, or responsibility toward their prey, changing how they see fish. There is a physiological impact when looking at or for a thing of passion because we see only what our brains subjectively choose. There is an acute focus in an oceanic literacy that ties together an act of awareness with specific sensations from the seascape that affect the body. While everyone might "see" the same visual cues, the Kanaka fisher's mind subjectively interprets and transforms information entering the brain through his ontological relationship to the ocean. She is engaging in a political act of seeing that can be discerned to know distinct worlds.

Smell and taste also become political sensations in seascape epistemology. As a Kanaka fisher smells and tastes the fish of the sea—two senses that connect directly to the hippocampus, the center of the brain's long-term memory—she is summoning recollections from ancestors. Feelings emerge as she smells the specific scent of fried *kole* (surgeonfish) as memories are stimulated of her grandmother teaching where and when these fish gathered in shallow waters to feed. As the fisher tastes the kole, not only does her mind recall how to catch and prepare this fish, but her body also vibrates as it remembers how to do so. All stimulants reaffirm her indigenous identity. The act of remembering alters constructions of the self. Like Marcel Proust's madeleine, the kole fish comes to occupy what the taste of the fish *means* to the Kanaka fisher rather than an occupation with taste as an isolated sensation (Lehrer 2007). It is the affective scent and flavor of fishes that help create pathways toward the remembrance of an autonomous and connected process of becoming for Kānaka.

Oceanic literacy also requires listening to waves formed thousands of miles away from the bowels of whirling winds. Surfers are trained to hear specific noises in nature that might sound like a cacophony of violently crashing water but which reveal distinct patterns of informative choruses when heard by an adept ear. I tune in to the familiar songs of the dominant ground swell, the most frequent guest along my home shores, noticing when this song becomes distorted or rearranged, indicating a change in waves. I can feel the sound vibrations of waves stamped across the ocean, traveling unimpeded by land until their liquid legs slow upon encountering the seabed, toppling their heads, still traveling rapidly, rolling them over as the wave curls into a perfect gift of mobility for fiberglass or wood boards, canoes, bodies, animals, plants, and spirits. Listening to the always changing and newly created voices of the sea through the ears and in the various quivering of the cells challenges the notion of what can be heard and how we hear. Oceanic literacy stimulates our minds to not only hear but also feel sound.

Similar to my ability to hear messages in unorthodox melodies, a surfer abstractly senses the changing shapes and contours of beaches by understanding the science of how waves break along them in seasonal sizes and varying directions. I rely upon an affective response to a lifetime of observations: impressions, sensations, and an embodied response to the constantly evolving rhythms of the sea. At Pipeline, one of the most challenging and popular surf breaks in the world, surfers survey the single geological structure of the rock that divides Pipeline from an adjacent break called Backdoor. Surfers know the niche at either side of the slab that comes to a point and runs at an angle to the beach. The northerly fissure creates the wave at Pipeline, and the southerly one, which is more distinct and shallow, forms the wave at Backdoor. Surfers read the boils, the water color, and the reef outline to adjust their positions and line up in the best spot to catch a wave, and they must also possess the ability to anticipate spatial movements and changes of wave locations. Surfers know that despite the reef location, the large winter swells will shift the sand surrounding the reef and beaches, affecting wave shape and location. Swell direction affects spatial movements. By learning to accept that what they have studied will inevitably change—erosion, tectonic plate shifting, coastal construction projects, modernization, and the mind—surfers are equipped to resist a single or narrow existence by fostering a way of knowing that is predicated upon the constant regeneration of the seascape.

The movements at Pipeline are paradoxical: it is the world's most famous surf destination and one of the most photographed waves, and a site consumed by the capitalist tourist project that depoliticizes and dehistoricizes

Kanaka identity and ʻāina, yet it is also part of the Hawaiian seascape, which is critical to the recovery of Kanaka identity and place. This surf break offers a space in which Hawaiian bodies experience ke kai with their muscles, senses, and memories in such a way that the surf repartitions and redistributes what can be seen and heard within the political and social structures of a tourist site. Heʻe nalu enables the body to redistribute movement through space with unconventional gestures that are both, and independently, interpreted by tourists and the Kanaka surfer.

The ligaments in the surfer's feet and ankles articulate a sliding across and above, vertically moving up and left. While this may be an expected sight for the tourist squinting out to sea, the knees of the Kanaka surfer, crouching under spitting tunnels of water, move her into a space of constant fluctuation that is untouchable by the dominant narrative of surf tourism: happy native playing in the water. The tourist site of Pipeline is also a regenerated space for re-creation as the Kanaka surfer's arms cut through polluted waters from runoff caused by tourist development. The ocean is experienced through bones and ligaments, enabling the surfer's identity to ride an embodied wave of potential while expressing a melodic symphony of diverse engagements.

This complex assembling of the ocean with the body, of reading and doing, is the political potential within oceanic literacy. Oceanic literacy helps to combine philosophical and physical interactions within and between the mind, body, and surrounding world. Knowledge speaks to the creation of an indigenous identity outside that of a citizen-subject, or as merely a "body" endowed with civil rights. The Kanaka navigator, fisher, and surfer are no longer predetermined identities within a rigid structure centered around conceptions of territory or written in a constitution, but are continually redefined through ongoing experiences with ke kai: seeing, smelling, and hearing is expanded to create alternative and autonomous realities.

Becoming the Voyage

"Seeing" ke kai enables a navigator, fisher, and surfer to move through it. Seascape epistemology becomes an embodied voyage for Kānaka Maoli, a way of knowing through corporeal interactions. This kinesthetic oceanic literacy, and its ability to re-create, reconnect, and empower, is illustrated through a historic Kanaka voyage: the Hawaiian canoe *Hōkūleʻa*'s 1980 sail to Tahiti.

The voyage began under a gray sky. Rain was falling from a thick mass of clouds and the wind was blowing out of the east, the worst direction to embark upon the 2,500-mile journey to the Tuamotu archipelago. The crew had waited for the weather to turn for five days. Waiting was difficult, however, because this inceptive voyage, while not Hōkūle'a's first to Tahiti, was to be the first with a Kanaka Maoli navigator and crew from Hawai'i. Unable to wait any longer, with poor winds, no sun, and choppy swells, on March 15 Hōkūle'a slowly glided out of Hilo Bay.

Nainoa Thompson was the young and determined Kanaka navigator of this voyage, eager to pull Tahiti out of the ocean, an achievement of great cultural and historical significance not performed by Kānaka Maoli for hundreds of years. After his experience as a crew member of the Hōkūle'a in 1976, Thompson saw the potential within the canoe to reawaken Kānaka 'Ōiwi to an empowering and intimate oceanic knowledge that was once an integral part of Hawaiian literacy. Thompson is an intense, intelligent, and quietly articulate man whose physical presence emits a kinesthetic wisdom. Always pensive and thoughtful, his body moves deliberately and eloquently, lightly and purposefully. Today, Thompson has become a cultural icon within the Hawaiian community, inspiring and helping to lead Kānaka Maoli toward cultural sovereignty. As he asserts, it is the ocean that has shaped who he is: "For decades of my life, I made an agreement that I needed to touch salt water everyday. It was a commitment to the ocean because it was my source of well-being and strength. And in making the commitment, I made myself stronger" (Thompson 2008). Thompson places himself in the sea, teaching the nerves in his eyes, his shoulders, and legs to feel its knowledge: listening, watching, reacting. Engaged with his surroundings, Thompson's body absorbs information while his mind imagines how to use it. This imagination is what helped Hōkūle'a to once again set sail in 1980, on a voyage that helped to reinvent a people.

The first days of the journey brought physical and emotional challenges: some of the crew got seasick from the rough conditions, and Thompson was consumed by an excited anxiety. Thompson knew that his finding Tahiti depended on his ability to communicate with the ocean, his capacity to sit himself inside its movements. Not only did his mind need to keep track of calculations, but his body needed to speak to the winds and waves. Thompson had to create an ocean-body assemblage under very physically and mentally challenging conditions.

Merging his body with the sea didn't come easily, he explained. It was the seascape that had to push him into this connected state by taking away all of his known points of reference. For Thompson, this pivotal point came when the wa'a entered the doldrums, an infamous zone of volatile weather. This is where Thompson's orthodox knowledge base was first challenged, the place where ke kai forced him to rely upon not only his mind and body for direction but also those internal sensations in his na'au (also known as "gut instinct").

As *Hōkūle'a* entered the doldrums, there was only darkness—no stars, no planets, staccato winds, switching waves. Typically, Kanaka navigators used to use ocean swells as guides rather than surface waves, which they knew changed direction with local wind shifts. Ocean swells, or ground swells, are generated by trade winds and distant storms, and move in a straight path from one house on the star path to another house of the same name on the opposite side of the horizon. For instance, a swell from the direction of Manu Ko'olau (northeast) will pass under the wa'a and head in the direction of Manu Kona (southeast); a swell from 'Āina Malanai (east southeast) will pass under the canoe and head in the direction of 'Āina Ho'olua (west northwest) (Polynesian Voyaging Society 2005a). Therefore, as Herb Kawainui Kāne, Kanaka artist, author, and cobuilder and first captain of the *Hōkūle'a*, explains, "Under overcast skies navigators could maintain their heading by keeping a constant ratio between the amount of pitching or rolling induced by a dominant swell, holding direction until guiding stars became visible again" (Kāne 1997, 24).

Yet in the doldrums, swells do not march in lines, they swirl and slap, making ground swells difficult to distinguish. And that night, the wa'a kaulua was traveling fast, about twenty-five knots, in an unknown direction. *Hōkūle'a*'s steersmen were asking their navigator for a direction that he didn't have. With no response, Thompson turned his back to them and walked away.

A navigator doesn't sleep much on a voyage, usually only a few hours each day, and even less when the conditions are rough. That night Thompson was exhausted. Bracing himself against the canoe's railing to avoid collapsing, he closed his eyes, flirting with sleep as he searched inside for a path. Under this oppressive exhaustion—arms limp, knees buckling, eyes swaying with the open ocean swells and orbiting stars—Thompson was able to release his analytical mind and allow his body to more freely intertwine itself with the seascape. This point of release was what enabled Thompson to suddenly "see" the rays of the moon hidden behind clouds.

No longer fighting to find it with his eyes, his body relaxed and he was able to feel its warmth on his right shoulder. "That's the truth, and that's the only way I can explain it," he says, still amazed by his experience (Thompson 2008). The moon was able to more clearly communicate with him, to effectively speak to his body through light sensations not seen by his eyes but felt by his skin.

Hitting a point of extreme emotional and physical exhaustion didn't inhibit Thompson's sight but expanded it. He was suddenly able to experience the world through new modes: "The most powerful times are not during arrivals, but during the journey. That's where I'm most raw and I'm most connected. So for me, somehow, that has shaped everything about who I am. And I've always been there, but voyaging gives me the opportunity to be able to understand and express it in a powerful way" (Thompson 2008).

That night Thompson came to understand himself as an integrated identity by merging with the seascape through an intensely affective experience that formed a kind of human counterpart to place (Hansen 2004). Thompson enacted a new potential of movement through a relationship with 'āina by relinquishing control. What Thompson could not visibly see in the doldrums initially frightened him in a way that did not instigate immobilization but stripped him of the logical security of "techniques" and "names," and enabled him to enter an alternative reality predicated on connection. He released his ego (not his identity, but the dissolving of any difference between the body and place) and was gifted a sensory insight into how the world functions.

Thompson has realized seascape epistemology by setting into motion an indigenous way of traveling that orients the body inside the moving kai as opposed to vectoring the self "off a point that intersects latitude and longitude as an absolute map of social and natural space," as stated by Vincente Diaz (Diaz 2009, 12). In Oceanic navigation, everything is connected and moving within the twelve thousand vast miles of Pacific Ocean, creating a guidance system anchored in relationships rather than abstract geographic categories. The fundamental concept is that stars mark islands and reefs not in relation to the Western cardinal directions of north, south, east, west, but in relation to other islands and reefs, all of which move in a synchronized pattern.

The stars, for instance, rotate in a star path, above islands and reefs also moving along these same paths. Pacific Islanders understand that the islands and reefs are mobile, constantly expanding and contracting with the movements of their indigenous inhabitants: the islands expand as the fish, birds, and plants travel out, and they contract as their inhabitants journey back home. The "boundaries" of the islands and reefs are perceived as fluid and connected to the life within them; this is the Carolinian concept of *pookof*. Within this indigenous understanding of space, the canoe is the stationary point as the islands flow past on the sea; this is the Carolinian concept of *etak*, which will be further explained in a moment. Polynesians traditionally used a similar concept. Kāne explains, "Whereas the modern navigator is equipped to fix his position without reference to his place of departure, Polynesian navigators used a system that was home-oriented. He kept a mental record of all courses steered and all phenomena affecting the movement of the canoe, tracing these backwards in his mind so that at any time he could point in the approximate direction of his home island and estimate the sailing time required to reach it—a complex feat of dead reckoning" (Kāne 1997, 20).

To voyage to Tahiti in 1980, Thompson needed to become the voyage: a personal, cultural, and political voyage. He needed to not only re-create how he understood himself as related to ke kai but also re-create the Hawaiian oceanic literacy of hoʻokele, which had been lost from colonization. Mentored by the late Micronesian master Mau Piailug, Thompson began adopting Oceanic navigational concepts as he watched what Piailug saw in the ocean, listened to what he heard, and felt what he touched. These lessons were then adapted to accommodate a specifically Hawaiian cultural and geographic "location" in Moana. Colonial history has, in part, required a contemporary Kanaka literacy to include both an indigenous and modern approach to knowing; a mixing of Hawaiian knowledge and Micronesian concepts, as well as Western technology and science. For instance, Thompson explains how he initially learned to measure the speed of a waʻa kaulua, a critical element in hoʻokele, as a way of discerning distance traveled: "I counted the time it takes the foam to get from the front ʻiako [outrigger boom] to the back ʻiako; that is a purely scientific, mathematical equation. That's not traditional. . . . I made that up because I didn't know how else to understand speed. At first, I had to stay back in that framework that I could trust and I was confident in. The transition to the navigation is when you say, 'I don't need that anymore'" (Thompson 2008).

Thompson had to allow his body to embrace the spatial and temporal rhythms of the seascape within his evolving mind. What he created is what he called a Kanaka "star compass" (see figure 4.2).[1] It is worth noting that while Thompson initially called his system a star "compass," because this

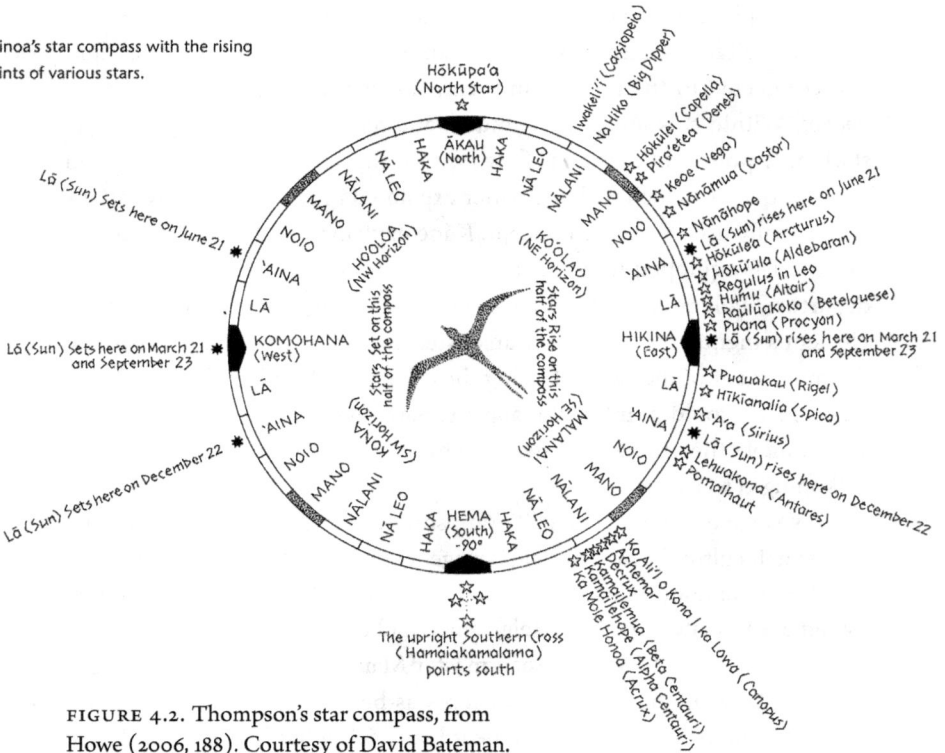

Nainoa's star compass with the rising points of various stars.

Hōkūpaʻa (North Star) ☆

ʻĀKAU (North)

HAKA • HAKA

NĀ LEO • NĀ LEO

NĀLANI • NĀLANI

MANO • MANO

NOIO • NOIO

ʻĀINA • ʻĀINA

LĀ • LĀ

HOʻOLOA (NW Horizon)

KOʻOLAU (NE Horizon)

Iwakeliʻi (Cassiopeia)
Nā Hiku (Big Dipper)
Hōkūlei (Capella)
Piraʻetea (Deneb)
Keoe (Vega)
Nānāmua (Castor)
Nānāhope
Lā (Sun) rises here on June 21
Hōkūleʻa (Arcturus)
Hōkūʻula (Aldebaran)
Regulus in Leo
Humu (Altair)
Raūlūakoko (Betelgeuse)
Puana (Procyon)
Lā (Sun) rises here on March 21 and September 23

Lā (Sun) Sets here on June 21 ✳

KOMOHANA (West)

Lā (Sun) Sets here on March 21 and September 23 ✳

Lā (Sun) Sets here on December 22 ✳

Stars set on this half of the compass

Stars Rise on this half of the compass

MALANAI (SE Horizon)

KONA (SW Horizon)

HIKINA (East)

Puauakau (Rigel)
Hīkianalia (Spica)
ʻAʻa (Sirius)
Lā (Sun) rises here on December 22
Lehuakona (Antares)
Pōmaihaut

HEMA (South) -90°

Ka Aliʻi o Kona / Ka Lawa / Ka Lawa (Canopus)
Achernar
Kamaʻilemua (Beta Centauri)
Kamaʻilehope (Alpha Centauri)
Ka Mole Honoa (Acrux)

☆ ☆
☆
The upright Southern Cross (Hamaiakamalama) points south

FIGURE 4.2. Thompson's star compass, from Howe (2006, 188). Courtesy of David Bateman.

is the term that Piailug used (Thompson notes that Piailug's English is not fluent, which may be why he chose this easily translatable term), his star compass erroneously infers a static grid of cardinal directions conceptualized by Europeans that fixes locations in the flexuous sea. The indigenous perception of Thompson's star compass is more accurately defined as a "path," a term he and other Kānaka Maoli also employ, because a "path" infers well-trodden lanes pushed through the sea that are rediscovered with each voyage through the seascape's changing patterns. The benefit in recognizing such semantic distinctions is to avoid unwittingly replicating the ways in which the West has shaped indigenous understandings of space and movement by blindly translating and glossing over indigenous seafaring cartography (Diaz 2009). Likewise, it is equally important to recognize the indigenous ability to use Western ideas and terms to articulate indigenous concepts by recreating them within an autonomous framework.

Thompson's star path becomes more than an oceanic guide; it becomes a cultural metaphor that Kānaka Maoli can use to navigate through life. For instance, at the top of the star path sits the North Star, or Hōkūpaʻa. Hōkūpaʻa means the "stuck" or "steadfast" star, named such because while the many other stars march across the night's sky, Hōkūpaʻa appears to remain nearly motionless in the northern hemisphere. Because its location is relatively constant, if a voyager can see this star, he can determine his direction and location beneath or in some angle to it (latitude). The Hōkūpaʻa might come to represent Hawaiian genealogy, a rooted origin that gives Kānaka location in the world. Like the Hōkūpaʻa, Hawaiian genealogy acts as a stable guide, even as Kānaka Maoli travel in time (modernity) and space (diaspora), adapting as complex individuals.

When I asked Thompson how a Kanaka navigator finds his location east or west of a destination (longitude), he responded that there is no absolute guide. Instead, he draws an imaginary line from his starting point to his destination point, and continually calculates how far east or west he travels off this line. Thompson delves into a conceptualization of space that enables movement through imagination, and which allows him to "know" where he is situated within a rotating world. His means of finding longitude, while distinct from the etak technique his mentor, is nonetheless built on the same concept. Etak, which translates as "moving islands," calculates distance traveled by fixing one's position within a triangle of three moving points (see figure 4.3). Diaz explains,

> First you steer towards the stars that mark the island of your destination. While doing so, you also back sight your island of departure until you can no longer see it. At the same time, you also calculate the rate at which a third island, off to the side, moves from beneath the stars where it sat when you left your island of departure, toward the stars under which it should sit if you were standing in the island of your destination. (Diaz 2009, 12)

David Lewis, author and navigational researcher, helps to further explain the concept of etak:

> Let us take the simplest case of a voyage that proceeds direct from island *A* to island *B*. A third island *C* is chosen as reference island.... Ideally it should be equidistant from the other two and located to one side of the line between them. In practice it is the exception to find one so conveniently sited.

The navigator knows how the reference island bears from island *A*. He also knows its bearing from *B*, it having been part of his training to learn the direction of every known island from every other one. In Carolinian terms he has learned "under which star" *C* lies when visualized from *A*. [see figure 4.3] . . . it lies under star *X*.

As the voyage moves toward the objective *B*, the bearing of island *C* alters until, when the canoe has reached the position shown in the diagram, *C* has come to lie beneath star *Y*, the next point of the sidereal compass. The canoe is then said to have traveled one *etak*, and this is expressed by saying that the *etak* island *C* has "moved" from one star point to the next, in this instance from under star *X* to under star *Y*.

This is the essence of the concept—that one *etak* along the course corresponds to the apparent movement backward by one star point of the reference island.

. . . The sea flows past and the island astern recedes while the destination comes nearer and the reference island moves back beneath the navigating stars until it comes abeam, and then moves on abaft the beam. (Lewis 1994, 174–75)

The concept in etak is that the canoe is the stationary point, while the sea and islands are mobile, traveling past the waʻa, hence "moving islands." The islands are pulled to the canoe as the stars and waves pass by. The seascape moves around this waʻa "center," allowing one to always know one's position and location as indexed by the signs in the natural and supernatural worlds in the surrounding seascape. The indigenous voyager becomes the center through an indigenous concept of time and space, allowing for an unambiguous sense of one's place within a fluidic pathway.

Yi-Fu Tuan states that the "'center' is not a particular point on the earth's surface; it is a concept in mythic thought rather than a deeply felt value bound to unique events and locality" (Tuan 1977, 150). The Pacific voyager becomes the center through a specific interaction and relationship with the seascape. His center is never fixed in physical location when voyaging; it is a location within his naʻau that guides him through ka moana. Created is an ideology of travel where one's roots can have routes, as Teresia Teaiwa might say. As mentioned, ʻāina, the Hawaiian Islands, remain crucial to Kanaka identity and thus political and social empowerment, but the *ideology* of voyaging includes the movement of these roots as Kānaka Maoli

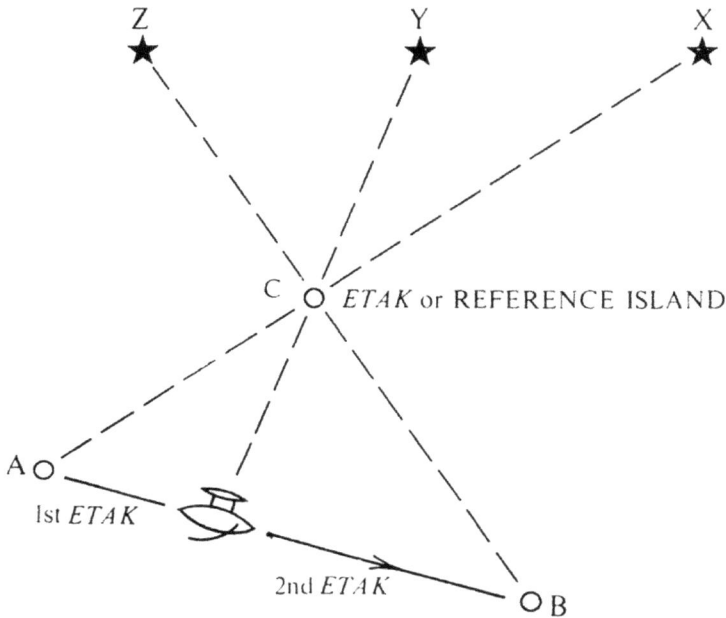

FIG. 30 A very short voyage of two *etak* from island A to island B

FIGURE 4.3. Diagram of *etak*, from Lewis (1994, 188). Courtesy of the University of Hawaiʻi Press.

voyage beyond the shores of their home. Thus, Kanaka voyagers lack an immediate reflex to ideologically impose their identities or culture as they travel, a concept explored in the following section. Instead, there is a focus on creating connections, rather than distinctions, between their traveling center and places encountered.

While Thompson doesn't technically employ etak in his own practice of hoʻokele (humbly insisting he doesn't completely understand the concept), the Kanaka moʻolelo of the demigod Maui illuminates how "moving islands" were also a part of Hawaiian epistemology. In the famous moʻolelo, "Maui The Fisherman," Maui fishes the Hawaiian Islands up out of the ocean, pulling them to his waʻa. The "Chant of Kualii" describes this event:

Oh the great fish hook of Maui!
Manai-i-ka-lani 'Made fast to the heavens'—
its name;

An earth-twisted cord ties the hook.
Engulfed from the lofty Kauiki.
Its bait the red billed Alae,
The bird made sacred to Hina.
It sinks far down to Hawaii,
Struggling and painfully dying.
Caught is the land under the water,
Floated up, up to the surface,
But Hina had a wing of the bird
And broke the land under the water.
Below, was the bait snatched away
And eaten at once by the fishes,
The Ulua of the deep muddy places.
(Westervelt 1910, 12)

While etak is not technically practiced, the concept remains critical for contemporary Kanaka movement beyond local activities so that Hawaiian identity can be diverse and mobile. In other words, while Hōkūpaʻa represents Kanaka travel through time in relation to genealogy and an origin, diasporic, philosophical, and conceptual movement in space requires a spatial conception akin to etak.

Thompson's star path engages a discourse of place as a dynamic composition. While structured around visual cues from the scientific and physically visible world, the star path ultimately relies on a "figurative" world in which to function. Carter explains,

> To figure something out means to think figuratively. It is to associate formerly distant things on the basis of some imagined likeness. It is to draw together things formerly remote from one another. The line of such thought represents a movement, a dynamic contraction that cannot be adequately represented by the dimensionless line of cartography. To think figuratively is to inhabit a different country of thought. In this, the ground cannot be taken for granted as a uniform and flat plane in which the ideal figures of thought are incised. The environment of figurative thinking possesses topological properties: it has points that lie far apart but belong together; it also has surfaces that look close together but in fact never meet. It is a world where the laws governing relationships count, and where the value of passages is recognized. (Carter 2009, 6)

The "value of passages" are central to Thompson's path, in which everything is pulled together through imagination. It is Thompson's imaginary line that draws islands together, creating a relationship rather than a gap. Identity is assembled within the movements of the ocean as ke kino moves through it, meshing the two together. Thompson says, "I don't really know what that [connection] is, but I know it because I've experienced it. I know it's true because it's happened" (Thompson 2008).

Voyaging Becomes an Ideology

Travel is a dynamic and specific form of movement capable of either solidifying or displacing cultures and identities. It allows for the possibility of new opportunities and empowerment, as well as for further oppression and marginalization (Diaz and Kauanui 2001). Travel is not a concept but an ambiguous apparatus, a way of thinking that is "at once enmeshed in the problematic of colonial power yet available to the analysis and critique of it," making possible both the disruption and continuation of colonialism (Thomas 1994, 4).

Travel, then, is much more than movement across space; it is movement across "location." James Clifford writes, "'Location,' here is not a matter of finding a stable 'home' or of discovering a common experience. Rather it is a matter of being aware of the difference that makes a difference in concrete situations, of recognizing the various inscriptions, 'places,' or 'histories' that both empower and inhibit the construction of theoretical categories like 'Woman,' 'Patriarchy,' or 'colonization,' categories essential to political action as well as to serious comparative knowledge" (Clifford 1989, 4).

Location is thus a series of encounters, traveling within diverse but limited spaces. Edward Said describes a traveler who ventures across these encounters: "The image of traveler depends not on power, but on motion, on a willingness to go into different worlds, use different idioms, and understand a variety of disguises, masks, and rhetorics. Travelers must suspend the claim of customary routine in order to live in new rhythms and rituals . . . the traveler *crosses over*, traverses territory, and abandons fixed positions all the time" (Said 1993, 6). And as travelers cross over locations, the act of travel itself modifies, informs, and even creates ideologies. Travel becomes an ideology that expands and contracts spaces or locations.

To clarify the concept of an ideology as employed by this work, I begin with a statement offered by David Bakhurst, professor specializing in philosophy and psychology, and concepts of the Self: "The term 'ideology' was coined by Destutt de Tracy in 1796 to refer to the 'essence of ideas.' This discipline, inspired by the EMPIRICISM of BACON, LOCKE and Condillac, was to give a NATURALISTIC explanation of the processes by which the mind forms thoughts" (Dancy and Sosa 1992, 191).

Ideology, then, can be better understood as both thinking processes and the results of these processes. It is not merely the body of ideas pertaining to, derived from, and reflecting the social needs and aspirations of an individual, group, class, or culture. Ideology is both how and what one thinks.

I say that travel "becomes" an ideology because although the word "travel" is a verb and an action, *The American Heritage Dictionary* also defines it as a process, "the act or process of traveling." Ideology is also a process; the process of traveling *becomes* the process by which the mind forms thoughts. Following this line of thinking and the definition of ideology above, the process of traveling becomes a process by which the mind forms a set of ideas that create a "map" of how the world is structured and should function. When viewed from the broader perspective used by Bakhurst, travel also becomes a process that provides a map of how power becomes allocated among groups and individuals within these structures and functions.

What becomes critical in the use of travel as an ideology is the ability to map epistemology or ways of thinking that have crystallized into power structures. This mapping reveals and thus has the potential to resist dominant ideologies that have naturalized, marginalized, and become seemingly eternalized through established realities and dominant systems. An ideology of travel then, is a set of ethical ideas that create a "map" of the way in which the world moves according to that particular ideology. Revealed are both the power structures inherent within the ideology as well as alternative locations between these structures.

Travel ideology is also pluralized through the implementation of disparate epistemologies. In other words, different applications of travel ideology reveal distinct political dynamics within it. Western travel practices embrace(d) an ideology of representation of places and spaces through binary oppositions that naturalize a specific political philosophy and order. The ocean was perceived as Other, justifying Westerners' desire to control and colonize both the seascape and those encountered within the seascape. Seascape epistemology, however, allows Kānaka Maoli to establish travel practices with a distinct ideology of travel from the dominant literature. This indigenous epistemology approaches the ocean and surrounding islands as imagined extensions of self, avoiding the creation of a colonial ideology founded in binary oppositions, such as us/them and civilized/uncivilized, predominantly seen in a Western travel ideology. Seascape epistemology demands a center that is always moving (the waʻa). It demands a travel ideology that slides, is flexible, and creates distinct ecological and social ethics for Kānaka Maoli. Revealed are the political dynamics of an indigenous epistemology evolving within and through the ideology of travel.

For example, in 1778, as Kānaka were still actively voyaging throughout Moana, captain James Cook embarked upon a very different type of ocean movement. Cook was among the first Europeans to apply the revolutionary mathematical, textual, and linear voyaging techniques discovered during the eighteenth century, which enabled Europeans to move away from the coastline and venture into a realm of longitude. Sailing vessels began to venture "out there" with the goal of "arrival" and "discovery" within what was still perceived as a chaotic and mysterious place colored by biblical and mythical images of sea monsters, voracious whales, and catastrophic floods. The book of Genesis presented the ocean as the "great abyss," what historian Alain Corbin calls "a place of unfathomable mysteries, an uncharted liquid mass, the image of the infinite and the unimaginable over which the Spirit of God moved at the dawn of Creation" (Corbin 1994, 1–2). The ocean was the remnant of the Flood, and its endless movement suggested the coming of another flood (although the apocalypse in the book of Revelation has it that the end will come by fire sent by God, the ocean's anger was expected to play a part at the beginning of the series of cataclysms). The sea was the unknowable to Europeans; in fact, "to attempt to fathom the mysteries of the ocean bordered on sacrilege, like an attempt to penetrate the impenetrable nature of God" (2).

This new ability to jump across the abyss altered Western perceptions of space into organized territories filled with political and economic aspirations. Cook's traveling instigated European discussions concerning humanity and power that supported various images of political legitimacy, ethics of engagement, and accounts of global justice. Philosophy and political science scholar Brian Richardson writes,

> The representations of the world in Cook's voyages have political implications. In Europe, the representations of the places in the South Pacific were used in debates over the limits and character of human nature, over the relationship between science and politics, and

over the legitimate use of power, both at home and throughout the world. . . . What Cook found on the islands of the South Pacific, in other words, was used to idealize the natural political order that nationalists spent the next two centuries trying to create. The specific representations of the geographical and natural world offered to the readers of Cook's voyages were used to rework problems in political philosophy. (Richardson 2005, 7–8)

Cook's voyages represent the Western notion of the existence and nonexistence of places, of the mapping and reinscription, management and reinforcement of colonial power. Cook "located" every significant and previously insignificant place in a static grid of coordinates marked across the earth, creating a cartogram of Western place and meaning making (7–8). This dividing and systematic mapping facilitated the categorization and control of space. Modern spatial divisions evolved that included racial and gender spatial divisions that established Hawaiʻi and the rest of Oceania as both empty and opportune for settlement.

Again, voyaging becomes an ideology, an ideology that can be used to map epistemology. Movement can liberate, and, as seen in Cook's cartography, it can also establish specific agendas and doctrines of control. The difference lies in the distinct forms of earth writing. In the West, the sea is where the land stops. Richardson writes, "As is the case with crossings of the Atlantic and the Pacific oceans [by Europeans], the gulf has very little space in the narrative. The void was jumped successfully, and the articulation of places could begin again once the farther shore was sighted" (25). It was the coastline, the edge of civilization, that was the fundamental reference point for locating "places" centered around the dominant land mass, the familiar, the self.

Yet coastlines do not have fixed dimensions. In fact, the Atlantic Ocean is growing, pushing apart the continents with the daily creation of new oceanic crust from the great ridge of volcanic mountains curving down the center of the Mid-Atlantic Range. The coastlines of the Atlantic continents are therefore growing, while those of the Pacific are being swallowed into the Earth's molten interior. The resulting crustal turbulence (earthquakes and volcanoes) cause rapid alterations and sometimes total disappearance of coastlines (Welland 2009). The coastline is mobile.

Without the concept of a stable coastline, however, Western travelers would not have been able to draw maps, nor would there be a way to complete or arrive from a journey. The "coast," while continually shifting, is nonetheless a present and tangible geographical object, most reliably found protruding up and out the sea. A "coastline," however, as Carter sees it, "represents the traveling cartographer's desire to establish 'general principles'" (Carter 2009, 50). Carter posits that coastlines are, in fact, an effort to bring systemization to the landscape of the world by connecting composed locations in a particular order. Coastlines were burned onto Western maps as an authoritative "style of earth writing" that mummified knowledge and intellectual movement. Paradoxically, early European travelers searching for new knowledge were erasing the world's complexity by strategically simplifying it into what Carter terms an "Enlightenment geographical discourse" (52).

For example, although the Greeks have traditionally connected the act of journeying with that of seeking knowledge, they also confined knowledge when voyaging into a Greek framework. Professor of Greek literature and culture, Carol Dougherty, writes in the introduction of her book *The Raft of Odysseus: The Ethnographic Imagination of Homer's Odyssey* that Solon, the sixth-century BCE Athenian lawgiver and one of the seven sages, traveled widely after establishing his new set of laws in Athens in part for the purpose of sightseeing, or *theoria* (Dougherty 2001, 3). Theoria, from which we derive our words "theory" and "theorize," designates the process of traveling as to see something, as well as to "speculate" or "think about." This established connection between travel and theory, as well as between travel and ideology, are reflected in the *Odyssey*, Homer's epic poem about a Greek hero's ten-year journey across the sea, his many encounters with strange cultures and places, and his long desired return back home to "civilization." Odysseus's travels conceptualize Western cultures' first experiences with "new" worlds, including perceptions of those worlds and cultures about the self and about the Other. His travels, as Dougherty notes, "traced the contours of Greek identity; he marks its boundaries" (6). In fact, Odysseus voyaged to the ultimate boundary of Greek experience, right to the point of no return: the Underworld, land of the Lotus-Eaters, and the island of Circe. The *Odyssey* is at the heart of the vision that Greeks have of themselves and of others: it has provided a "long-term paradigm" for seeing and talking about the world, for traveling it, and representing it. The poem establishes a model for living in the world and for making it human, or Greek (6).

Odysseus's sea travels do not imply that "non-Greek" is set in binary opposition to "Greek," although there was a clear distinction between "home" and "abroad," between "land" and "sea," between "us" and "them." His travels at sea do, however, establish that the Greeks' sense of Other is reflected in their sense of self. There is a return to the self through a journey from the Other (15). Thus not only is travel "out there" critical to acquiring knowledge, but so is the "return home," the return to the self. This return allows the foreign to be understood, but only through the concept of the self. An alternative would be to travel not in order to revisit the "there" of one's identity, to not impose one's agenda or definitions upon the spaces and places through which one travels. An alternative would be to allow the self to travel inside spaces and places, inside the waves and islands, rather than have them travel through you (Benítez-Rojo 1996). Voyaging in this way allows one to escape the pitfall of idealizing, myth making, and depoliticizing the Other with ideas of "paradise" and desires.

Odysseus's need to "return" reveals the political and ideological ways in which Western mimesis traveled and was mapped. Greek colonial movement in the Mediterranean, which involved a fear of losing the self as well as a desire to discover the Other, followed Odysseus's dichotomous logic: he explored new territory while surviving the strange and new places of encounter in the new world. Whom and what are encountered within the space and time of the seascape created an imagination of difference for the Greeks as they employed governing motifs of ethnographic and anthropological discourses that informed colonial expansion.

The *Odyssey* exemplifies Carter's "Enlightenment geographical discourse" established by European travelers, cartographers, and thinkers blind to the mobility of the map's line and to the movement of the body thrown into spaces in-between, a "throwing" into the map that, Carter argues, draws the world together rather than separates it. Nature is statistically established as a machine that can be disassembled and controlled, a way of thinking enhanced by the Renaissance and Enlightenment, which spurred the linear measurement of time and space in the seventeenth and eighteenth centuries. This measuring divided nations on paper, encouraging nationalist sentiments and binary oppositions that governed spatial constructions of the environment as well as social and political relations between cultures. An analysis of how people voyage(d) disrobes an exemplary role of cartography in the demonstration of colonial discursive practices. How we move affects how identities and places are made; it changes how the world exists.

It should be noted that both Euro-settler and Pacific voyaging both took and take part in the production of specific politico-cultural orders that often articulate narrow horizons of identity, nation, participation, and government. Both Western and indigenous voyagers establish(ed) the dichotomy between home and away. Every center is someone else's periphery, marginalizing those outside of the dominant theoretical landscape (Clifford 1989). Travel ideologies should not be uniformly demonized or romanticized; both European and Oceanic travelers participated in a quest for new knowledge as well as conquest and oppression. Motivations for early Hawaiian voyaging were manifold: exploration, adventure, the discovery of new lands and resources because of growing populations or prolonged droughts or other ecological disasters in homelands, waging war, seeking vengeance, escape of political persecution, unhappy love affairs, visiting relatives, gift exchanging, tattooing, or the seeking of prized objects, such as red feathers, not found in the Hawaiian Islands (Kawaharada 1999).

Yet the Oceanic maps that guided Pacific Islanders to the islands they settled were inherently mobile. From a rooted genealogy, movement is privileged within the map, a way of constructing the world that allows, and even requires, the islands and stars to participate in the voyage. For Pacific Islanders, travel, and even settlement, is an interaction rather a reduction of thoughts into an "algebra of points and lines" (Carter 2009, 8). In Pacific navigation, the map moves with the voyager. Ancestors are always with Kānaka Maoli on the sea: ke akua Kāne is in every rain cloud and flash of lightning, and Kanaloa is in the waves and seashells. In this way, hoʻokele becomes a critical act of recovery for contemporary Kānaka that can resist imposed compositions of identity and place that feed an embryonic cycle of established "rights" and "property" while privileging an indigenous relationship to and understanding of the world through Hawaiian genealogical experiences of living in it.

Imagining the Voyage

When navigating, Thompson's body is soaked by the seascape. Each puff of wind and slight change in humidity moves through him. Thompson's oceanic literacy is accumulated into an experiential reserve so that, over time, he "begin[s] to anticipate the weather. It is like being inside the navigation, participating from the inside" (Low 2008). Thompson exists inside movement itself, as movement becomes a mode of experience transferred from place to body. Thompson's eyes, hands, back, and heart become critical sources of meaning and the precondition for communication (Hansen 2004). His affective experience in the ocean introduces the power of creativity into the sensorimotor body, enabling an expression of imagination through kinesthetic movements.

Thompson imagines this connection between kino and kai, this metamorphosis of awareness and embodied ability to read the seascape. To "imagine" is a verb that may conjure connotations of the unreal, fantastical, and thus unqualified in "reality." Imagination, however, generates knowledge by engaging the necessary process of composing concepts. In ʻŌlelo Hawaiʻi, hoʻomoeā means "to imagine deliberately." This is what Thompson is doing; he is deliberately creating a route of travel. He must hoʻomoeā an island before embarking upon the thousand-mile voyage toward it. It is his imagination that realizes pathways that often slide between dominant and preestablished lines and boundaries, toward alternative destinations. Thompson's mind is opened to visions not habitual to the mainstream brain. In this way, engaging the literacy of hoʻokele enables heterogeneity.

Yet imagination doesn't float aimlessly when voyaging. Thompson's visions are anchored in a specific cultural context about connection to place, which must be tapped into when clouds bury the skies and currents confuse swells. There must be something to draw upon when there is nothing material to see, hear, smell, taste, or touch in the seascape. Thompson's ability to "see" in the darkness of the doldrums upon that 1980 voyage to Tahiti is mana. This divine power is expressed through connections to ancestors, plants, and skies, but mana also requires the ability to act upon these connections for empowerment. Mana is what enables a navigator to hear the secretes in the sea, and to enter into an intuitive state of being, which Thompson calls the "zone": "There's a zone you get into in navigation; that confidence, that level of mana, that knowing without knowing at all how you know. And that's awesome. That to me is freedom. It's intellectual, physical freedom" (Thompson 2008).

The zone is a space and time of cultural relationship in which the body is completely affected by the sea. Thompson continues, "No matter what the conditions, everything is connected. That means the heavens, the clouds, the wind, and the waves, they're all connected, they're all making sense. When you're not in the zone, the pieces are disconnected and they don't make sense. . . . There's no class you can go take to get in the zone. That's a learned, educated, personal step. . . . If you don't believe it, it ain't going to happen" (Thompson 2008).

If you don't hoʻomoeā the connection, it won't exist. A navigator's main responsibility is to find his mana, to enter this zone inside himself. He must pull the seascape into his identity and existence, into his thoughts and movements, to create a dialogue that includes the thoughts and movements of the kai and of the waʻa.

Blankenfeld understands that the canoe is also a living entity built with the mana from the trees and hands of skilled Kānaka Maoli, and must not only be respected but used for travel: "When you sail across the ocean *with* the Hōkūleʻa that's what you're doing, you're sailing *with* Hōkūleʻa. She is like a spiritual entity with mana. There's no doubt about that. You're not sailing on Hōkūleʻa, you're sailing with her. It's a whole different thing, and of course as a navigator you're guiding her and you're directing with the crew, but there is an intimate connection there. It's always been like that, within Polynesia, it's always been like that" (Blankenfeld 2008).

In the same way, a Kanaka navigator sails *with* the seascape, not on it, but as part of it. Thompson recalls a story about how Piailug's grandfather taught him to weave himself into the sea so that he could voyage through it: "At age five, Mau's sailing. He's on the voyaging canoe, gets seasick, so his grandfather ties his hands together with sennit, throws him overboard, and drags him behind the canoe. Mau said to me, 'You know, when I was on the canoe and the waves made me sick, my grandfather ties me by the hands and throws me in the wave, so that I become the wave'" (Thompson 2008).

To be truly connected to the seascape, Piailug had to be completely immersed in it, even overpowered by it. He had to renounce control to understand how to move inside the wave. His body and mind were, and are, defined in terms of a larger totality: places, bodies, ideas, sharks, salts, rhythms, improvisations. This does not mean that Piailug is determined by the seascape, but he is instead shaped through an engagement with it, an engagement that is physiological, intellectual, and ancestral.

A navigator engages his body, mind, *and* ancestors to sail because, as Thompson says, there is no rational or scientific way to explain the magical things that happen on the waʻa, such as "seeing" the moon on the 1980 voyage to Tahiti. But these experiences aren't frequent for him; Thompson needs to immerse himself into an alternative realm of mana, a spiritual realm, which can take days for him to reach. Thompson asserts, however, that Blankenfeld can enter this world within minutes. "Ancestral" becomes the only word Thompson can use to describe the oceanic literacy Blankenfeld possesses: "I don't know any other way to say this, but I believe there is something very deep about an ocean ancestry that Bruce has, that we don't know, but that comes from some deep place and ancestry. He is ʻohana, he's part of that family that comes down. I believe that because he's so extraordinary, there's no one like him, you know" (Blankenfeld 2008).

The inference is not that *knowledge* about the ocean is genealogical for Blankenfeld or any Native Hawaiian. Kanaka oceanic literacy is not spoken of in Lamarckian terms; knowledge is not inherited.[2] A "genealogical" element of knowledge instead indicates that the practice of ocean knowledge holds a specific cultural significance for Kānaka Maoli because Hawaiian genealogies and identities are forever entangled with the ocean. Kanaka stories are submerged just under the sea's cellular membrane, rippling and following the patterns of the ocean's waves marching toward shore. Genealogy of this oceanic knowledge is not transferred or channeled from but digested in a way that cannot be rationalized because it involves a merging and fusing of mathematics, physics, dreams, and vibrations, but it is a connection to knowledge that has been experienced and continues to be articulated through oral histories.

There is a feeling of connection to ancestors who engaged the same sea when voyaging which, Thompson explains, "is culturally genetic. That is where I come from, my Hawaiian side is from the ocean" (Thompson 2008). This indigenous knowledge involves a depth of culture, an accumulated bank of insight that is not inherent, but that sits on the skin, and it rubs off on subsequent generations who pay attention to this established relationship. Blankenfeld's DNA has a memory; the genes passed down to him from his ancestors carry the quivering sensations of oceanic literacy inside them, which does not *transfer* knowledge but resonates with memories that can be accessed through instinctive attuning. These rhythms survive like floating notes in the sea that can be fished out through seascape epistemology.

Voyaging into the Future

As navigators, Piailug, Thompson, and Blankenfeld have not achieved anything novel or enlightened. They have, however, put into practice an art and an ethics by the re-creation and reenactment of something historically empowering. The art is the literacy, and the ethics is the orientation this literacy engages. Pacific navigators sustain a relationship to ka moana based on the concept of kuleana, to the 'āina. Kuleana is a right and entitlement bestowed upon the indigenous peoples of the Pacific Islands to carry the knowledge and memories within the seascape. The ocean's secrets are like shadows on a shark's back until they are envisioned by a voyager whose cracked hands harness its breath with the pull of a rope. Immersed in an imagination of place, Kānaka Maoli are equipped with the sight and sensations necessary to extract the ocean's transcendental dimensions. Hawaiian kuleana to the ocean involves this interaction: connecting, remembering, engaging.

All of humanity has a kuleana to the ocean. The salty sea touches everyone in one form or another: the droplets of water lapping on one shore are soon over the horizon, wash up on another's beach, or evaporate into the clouds that blow over and rain onto another's valley floor. The knowledge and memories in the sea are not ours to lose; they belong to our species as gifts from our ancestors to future generations. In this way, the oceanic literacy of hoʻokele is performed not only for contemporary Kānaka Maoli but for all of the world's children not yet born. This is the vision Thompson holds for the waʻa kaulua; it is a tool capable of instigating change: "If we agree we need to save the ocean, if we agree that Hawaiʻi is special, and we agree that there needs to be new kinds of education, then let's agree that none of us know the answer by ourselves. Let's agree that none of us have the path. None of us have the breadth and the experiences and the intelligence to know how to shift the world. But collectively we do" (Thompson 2008).

Creating this global shift is his kuleana. Having been gifted the oceanic literacy of hoʻokele by Piailug, his crew, his ancestors, and the ocean itself, Thompson considers himself the "luckiest person in the world." Every voyage he embarks upon, and with every expression of his literacy, the large paradoxes in life become small epiphanies. The performance of hoʻokele continually illuminates his "location" in the world, affirming where he came from and where he hopes to go. Oceanic literacy has the potential to transform. Now it is his duty and his honor, as he says, to continue to gift this knowledge to future generations. That, Thompson believes, is why *Hōkūleʻa* continues to voyage today, to construct identities intertwined with the ocean. From this ocean-body assemblage, people learn to love and care for this immense and life-sustaining sea, which in turn bestows life and wisdom upon us.

Now Thompson is sailing *Hōkūleʻa* around the world (which will be discussed further in chapter five). This four-year worldwide voyage, called Mālama Honua (Caring for Our World), he believes, will help to create a collective consciousness. "Extraordinary things are done by connecting extraordinary people," he proclaims (Thompson 2008). He believes that the indigenous vessel of *Hōkūleʻa* has the potential to carry a message of peace and environmental awareness, acting like a magnet for critical and innovative thinking. The hope is for the waʻa kaulua to create a physical and intellectual bond between mindful people, pulling the world together through oceanic movement.

Seascape epistemology is this movement, but it is also what is found in-between. It is the gap itself (Carter 2009, 98). Seascape epistemology is the note left out by Thelonious Monk, which does not silence music but creates an expressive style of jazz. Seascape epistemology is this interplay of movement and stasis that Thompson hopes the Mālama Honua voyage will realize: a specific cultural movement that doesn't magnify difference but connects through and because of it. Hoʻokele, as a specifically Hawaiian literacy, does not exclude; it inspires and connects. Mālama Honua will be an event that suspends the dialectic in favor of possibility. Movement does not occur in a single mode or gesture; how one moves through spaces and places reflects a larger political and ethical system. The fluidity of seascape epistemology enables knowledge to be resilient and multifarious, to be adaptable while maintaining a rooted source of cultural history. Knowledge becomes an emerging process, born from the movements of bodies affected by places, which can be engaged for specific purposes, only to be readjusted by more movement.

KA HĀLAU O KE KAI Potential Applications of Seascape Epistemology

One means by which seascape epistemology can be realized and applied for contemporary Kānaka ʻŌiwi in Hawaiʻi is through the concept of *ka hālau o ke kai* (the ocean gathering house) as an educational center that teaches, applies, and supports oceanic literacy: hoʻokele, heʻe nalu, lawaiʻa, limu picking, and reef and ecological care, as well as knowledge about the marine life and environment within the particular ahupuaʻa in which the hālau is located.[1] The entire ahupuaʻa is emphasized in ka hālau, upholding the Kanaka concept of ʻāina as incorporating both land and sea. Ka hālau is a place to realize the ideas, concepts, and theories of the academe (such as Epeli Hauʻofa's "sea of islands") and a place where community culture, knowledge, and identity can be directly addressed and fostered. Bringing together academic strengths and abilities with those of Kanaka experience and culture creates a very powerful potential for Hawaiʻi. Ka hālau o ke kai also offers a space in which to engage the imagination for diverse and alternative futures prided in cultural confidence. It is an educational hub and research site, as well as a cultural center for the youth and people of Hawaiʻi.

Ka hālau is also a site that expands indigenous literacy, a modern "reading" and "writing" of literature, the environment, senses, people, and genealogy through diverse interactions, oral histories, songs, poems, dance, art, writing, and ceremony. The Western and dominant definition of literacy excludes these indigenous practices of reading the sea and of chanting one's genealogy. In response, ka hālau o ke kai expands the notion of literacy within both formal and informal systems of education. It is a place where the current and dominant system of education is disarticulated away from a Western-based system historically focused on the process of industrialization, and rearticulated toward an indigenous approach to learning, being, and knowing rooted in a relationship to ʻāina. (Re)connecting to seascape through an oceanic literacy helps reaffirm an indigenous and alternative identity by (re)connecting Kānaka Maoli to the natural elements so crucial to our identity. Ka hālau affirms cultural sovereignty and revitalizes and puts into action local ways of thinking and producing knowledge.

Ka hālau o ke kai becomes the bridge not only between Kānaka Maoli and our oceanic literacies but also the temporal disjuncture between very old and very new things, between the ancestral and modern (Clifford 2000, 96). Ka hālau reaches back to reading ka nā nalu, building waʻa, memorizing wind names and patterns, and seeing islands thousands of miles away, and it emphasizes oral education practices while making these literacies significant and applicable to Kānaka today through modern materials, technologies, modes of traveling, sense of place, recreation, and individual expression. Kānaka Maoli can be modern by reaffirming roots and connecting an indigenous epistemology to contemporary applications. The hālau redefines education for contemporary Kānaka, and redirects it toward a core that emphasizes concepts such as mālama ʻāina and kuleana, caring and responsibility for the seascape and the earth. It is a place where the experiential and sensational knowledge within the archive of oceanic literacy can be accessed and read toward the application of seascape epistemology.

An exploration of the ways in which ka hālau could apply the specific oceanic literacies of hoʻokele, heʻe nalu, and lawaiʻa, reveals the significance of each practice- and place-based literacy for contemporary Kānaka ʻŌiwi.[2] I suggest that ka hālau build upon the efforts of the voyaging waʻa kaulua *Hōkūleʻa*, the heʻe nalu clinic "Nā Kama Kai," and the fish pond at Heʻeia on Oʻahu, all of which provide living illustrations of how Kānaka Maoli engage oceanic literacy for cultural and political empowerment, a movement upon which ka hālau will build.

I do want to recognize that Kanaka organizations and schools are currently engaging such place- and practice-based educational systems and programs through cultural groups as well as charter and immersion schools.

Ka hālau o ke kai, however, would be the first physical gathering place for community ocean practices and education. Ka hālau also differs from immersion schools under the umbrella Pūnana Leo preschools and kula kaiapuni elementary schools, for example, as well as charter schools such as Hakipuʻu Kāneʻohe, Hālau Kū Māna in Mānoa, and Ka ʻUmeke Kāʻeo in Hilo, in that it is a school *and* community center focused entirely on the ocean. Ka hālau o ke kai would hope to work with and include these schools through outreach programs during and after school. The vision is to immerse students in oceanic literacy, but to also include the larger community by expanding the role of the hālau to become a gathering place for students and their families after school and on weekends, particularly for those living within the ahupuaʻa in which the hālau is situated. The result is the strengthening of community support by creating stewards of their own resources. Ka hālau would be a place for diversity and incorporation that emphasizes a Kanaka ancestral foundation within ke kai. It is a place to include a Kanaka narrative and to reassert Kanaka identities into knowledge production and practice in Hawaiʻi.

Ka Hālau o Ke Kai

Ka hālau o ke kai would consist of a set of several *hales* (houses) that sit along a coastline, possibly in the Kaloko region along the Ka Iwi coast on Oʻahu's southeastern shore historically in the *moku* (district) of Koʻolaupoko, and in the ʻili (subdivision of an ahupuaʻa) of Maunalua. Today, however, Kaloko is considered part of the moku of Kona, which was officially named Honolulu in 1859. The Ka Iwi coast region that I suggest runs between Makapuʻu Beach and Sandy Beach. Although the coast is rough with lava shelves and scrubby vegetation, Kaloko is an ideal area for several reasons: it has several historical fishponds available for restoration (*ka loko* means "the pond"); there are natural springs in the region; the far eastern-facing point of the region offers a good location for studying the weather; it is a naturally protected area due to lack of access from both the land and ocean; marine life is abundant in the area; and there are numerous tide pools for exploration. Kaloko is currently a state conservation district. Replanting native plants and caring for the marsh in this area would be wonderful projects for the hālau. The isolation also adds to the area's attraction, although increasing numbers of tourists are venturing into this region following suggestions of tour guides and books. This area remains ideal, however, because it is adjacent to the Waimānalo ahupuaʻa in the moku of Koʻolaupoko, which is home to a large Kanaka Maoli population, offering a rich source for *kūpuna* (grandparent or of the grandparent's generation) and students. The ahupuaʻa of Waimānalo also includes beaches more appropriate for launching canoes and for beginner surfing and diving, because although Kaloko has small coves and beaches, one being Kaloko Beach, now called Queen's Beach in Wāwāmalu Beach Park, the ideal beach for heʻe nalu, swimming, and launching waʻa is at Kumu Cove, or Baby Makapuʻu, just northeast of Makapuʻu Beach in Waimanalo.

In ka hālau, students and kūpuna would meet in one large, central hale that sits on the ahupuaʻa. *Lauhala* (pandanus) mats would line the floor of the main hale for sitting on, and the southern facing wall of the hale has a kapa, or bark cloth, acting as its wall, which could be risen for ventilation and viewing the ocean, or drawn down for privacy and warmth during cool evenings. There would be a big stainless steel kitchen in the hale for food preparation and storage. This hale would be used for general lessons in plant care, marine life recognition, hoʻokele, making ʻupena (fishing nets), the physics behind wave formation and riding, and fishhook carving, and for instruction on building fishponds and reading weather: currents, waves, clouds, wind, the moon, and so on. This hale would act as the main meeting center for ka hālau, where students report for daily instruction.

Outside the big hale would be bathrooms with lockers and changing rooms. There would also be outdoor showers with gutters draining water into nearby gardens. Only native plants would grow on the site, such as the *hau* (lowland tree), *kukui* (candlenut), *ulu* (breadfruit), and koa trees, all used for something related to the ocean: hau bark for making *kio* (rope); kukui nut meat to clear the water to see better when fishing, and kukui bark to dye finished fishing nets; and ulu and koa logs to make canoes, paddles, and surfboards. In this way, the ocean would be seen to be an extension of the land. Both are ʻāina. If the land is unhealthy, so is the ocean—just as the health of the sea affects that of the land.

There would be smaller hales surrounding the site, one of which would be for canoe building. This hale would be constructed from traditional materials and in a traditional architectural style, and would be built by students and community volunteers as part of the educational curriculum. The hale's maintenance would also be part of the curriculum. Another nearby hale would be for carving paddles; another for surfboard shaping; and still others for making ʻupena and fish traps.

Kūpuna would come to ka hālau to gather students and spread their knowledge before it is lost forever. Ka hālau would not only be for students; it would also offer kūpuna the opportunity to share their *mana'o* (thoughts or knowledge) and experiences; it would nourish a reciprocal relationship. Instruction would include oral, written, and active practices so as to engage all intellectual and physical aspects of knowing and expression. The *kumu* (source of knowledge) would initiate the student into certain Kanaka attitudes and ways of thinking. The fisherman, for instance, would introduce his or her student to a special relationship with the sea and its inhabitants. In the transmission of these skills and attitudes, nonverbal factors and means, such as mood, timing, setting, example, and imitation, are as important as verbal instruction and explanation. Oceanic literacy would be taught in both formal academic systems within ka hālau as well as through informal structures. In this way, seascape epistemology could survive as a theoretical philosophy about knowledge in the academe, illuminating the critical relationship between Kanaka culture and Kanaka environment, while also helping to perpetuate the practice of Kanaka culture in Kanaka spaces.

The relocation of language is an obvious and critical element of ka hālau o ke kai. Although 'Ōlelo Hawai'i, Hawaiian language, would not be the focus, Hawaiian language classes would be incorporated into the overall function of ka hālau. Hawaiian vocabulary would be used to explain concepts, processes, and the names of shells, fish, plants, places on the reef, places outside of the surf, different sand formations, and the star constellations. As 'Ōlelo Hawai'i is increasingly (re)learned, the hope is that ka hālau would encourage a linguistic resurgence in the community and in public spaces so that when engaging in a Kanaka activity such as he'e nalu or net fishing, 'Ōlelo Hawai'i will naturally express and support these activities.

Students would learn to perceive objects as natural phenomena through a Kanaka epistemology and ontology, using Hawaiian vocabulary and phrases to express knowledge and relationships, enabling access to and expanding the archive of oceanic literacy. For instance, as non-Kanaka Hawaiian scholar John Charlot explains, students are taken to the shore of the ocean and asked how many colors they see: "The Hawaiian words for different colors of ocean water are given as students see certain colors. Vocabulary also guides them to see certain colors they could not at first distinguish, for instance, *lipo*, the color of the ocean at the place the water is deepest. The students will thus realize that they are learning to see things they did not see before" (Charlot 2005, 30).

Students would begin to look at the ocean through Kanaka ways of knowing. This same method could be used for cloud colors and formations, coral and sand formations, marine life, wind directions, and seasons. Students would be connected to mo'olelo such as *The Wind Gourd of La'amaomao*, in which the protagonist, Pāka'a, must memorize all of the names of the wind in a chant in order to use the power of Lono (the god said to embody the gourd) so that Pāka'a can successfully and properly carry out the duties of leadership in Hawai'i.

The overarching goal of ka hālau is to physically get youth into ke kai—to smell it, hear it, taste it, touch, and experience it. Routine interaction with place provides a subconscious and inherent education unable to be taught through any other means. Students might not realize the breadth of the concepts, intuition, and relationships they would be naturally building by sitting at the shore and absorbing the ocean's dynamic and organic communications. It is my belief, however, that over time, students would realize how this interaction and cultivated relationship contributes to not only their identities and sense of self-sufficiency as Kānaka Maoli and human beings but also to their future life goals, ambitions, and careers regardless of focus or "discipline." The development of a seascape epistemology through oceanic literacy at ka hālau is a philosophy of knowledge and literacy that would transcend boundaries between land and sea, as well as divisive categories of "knowledge."

Ka hālau would be a place for young people to gather and engage in a Kanaka knowledge and to (re)connect with the 'āina through surfing, paddling, sailing, navigating, fishing, and diving. Classes would also incorporate related activities such as cleaning fish, drying fish, planting native plants, sewing nets, canoe surfing, and board shaping. I believe this system would successfully engage youth because there would be a sense of ownership of ka hālau and ka 'āina on which the many hale sit. Students would embrace the responsibility of caring for ka hālau, using their newly learned knowledge and skills, because they would have an investment in ka hālau, they would have helped build and create it as part of their own community. Furthermore, ka hālau would be a place for families to convene while engaging in cultural recreation and education. Parents would find a safe and positive place to leave their children, or actively participate with their children.

Practice-Based and Place-Based Education

Epeli Hau'ofa envisioned and helped to establish the Oceania Centre for Arts and Culture in 1997 to foster and nourish cultural creativity rooted in history, tradition, and an adaptation to modernization. The center is a place that encourages Pacific Islander development through imagination and sharing, emphasizing practices as the means by which issues of culture and identity are addressed. Its primary purpose is to encourage a flourishing of contemporary visual and performing arts through practice. Houston Wood, professor of English at Hawai'i Pacific University, explains, "At the Oceania Centre for Arts and Culture, Hau'ofa and his colleagues are not much interested in either the disciplines of art history and instruction or in formal interpretations of how their work fits into Pacific traditions. People at the Oceania Centre instead learn mostly through [Hau'ofa's words] 'observation and hands-on experience'" (Wood 2006, 42–43). Approaching learning and education through such an emphasis on practice-based knowledges promotes diversity in Oceania while supporting place-based autonomies.

Wood defines "practice-based research" as activities-focused research: "Practices are patterned activities that can be recognized as normal and repeatable by the people who enact them. Practice-based research thus may focus on any practices found along a continuum from the informal to the formal" (43). Practice-based education as defined by this work is based on Wood's definition of research; it is a means of learning and teaching based on patterned activities that can be recognized as normal and repeatable by the people who enact them through repetitive and experiential actions. It is a form of education based in both cognitive and embodied knowledges, on oral and active instruction of both formal and informal practices. In this sense, ka hālau o ke kai would engage formal rituals and protocols of, for instance, *pule* (prayer) before entering the classroom or before cutting down a tree to carve canoe paddles, as well as informal knowledges such as using sand to roughen the wax on a surfboard for improved grip.

Place-based education works in tandem with practice-based education in ka hālau o ke kai. Place-based education is a way of learning and teaching dependent upon a situated space that holds cultural, philosophical, and historical significance to those people engaging it, and that informs the practices that occur in that particular place. Davianna Pōmaika'i McGregor illustrates the relevance of place-based education in Kanaka society as she describes the relationship between Kuaʻāina and the ʻāina. *Kuaʻāina* literally translates as "back land," but came to refer to those Kānaka Maoli living the Hawaiian culture. Through the illustration of Kuaʻāina, McGregor portrays Kanaka ways of learning and knowing through place that are significant to social, economic, spiritual, and philosophical ways of surviving in and engaging the world:

> Kuaʻāina were intimately conscious of economic activities around the life cycles of the various fish, animals, and plants they depended upon for food. Thus, from month to month, as the seasons shifted from wet to dry, their food sources changed in accordance with the type of fish, fruits, and plants that were in season. This knowledge of the environment and natural life forces was often passed on and remembered as Native Hawaiian traditions and beliefs. Native Hawaiians often chose to personify the forces of nature as spiritual or akua and ʻaumākua, Gods and ancestral spirits. They created legends and myths to describe and remember the dynamic patterns of change that they observed. (McGregor 2007, 12–13)

For Kānaka Maoli, relationship with place is instrumental and fundamental in developing the basic principles from which all knowledge flows.

Practice- and place-based education is an approach to learning that crosses rigid disciplines that tend to separate the spiritual from the political, and activities from academics. The discursive practices of the Western-based system of knowledge and education places non–Western-based systems of knowledge outside the established knowledge and truth regime. It subjects Kānaka ʻŌiwi and other indigenous peoples to Western epistemological and theoretical approaches to education and knowing. The implication becomes one in which all peoples can be understood and can flourish using one system of thinking regardless of history, culture, or place.

Within this universalizing frame, University of Hawaiʻi professor Vilsoni Hereniko notes that the written word fixes the truth: "Genealogies, land titles, customary practices, secret rituals, disputes, religious beliefs (and so on) that were previously embedded in social relations are no longer subject to change or modification" (Hereniko 1999, 84). The notion is that there is one truth, based on written research that excludes emotional truth and the evolution of oral histories. Hereniko quotes a Fijian elder who states, "People do not understand the unseen, which is the reality of our lives; they do not realize its power. They look only at the seen, which is an illusion" (85).

Ka hālau o ke kai embraces diversity in an approach to learning, understanding, and relating to the world by emphasizing the oral, written, and practice- and place-based education of oceanic literacy. Expanding how we learn and interact with the world helps to make "truth" flexible and negotiable, as perceived by Hereniko. In fact, in a seascape epistemology, there are no final truths, just interpretations and negotiations. Practice- and place-based education allow for both oral and written words, for both rational and emotional theory, for both cognitive and experiential study. Wood asserts, "Those who emphasize practices generally reject the assumption that people live within self-organizing 'systems' of beliefs, values, norms, and symbols" (Wood 2006, 44). Encouraged is a comingling of contemporary and historical knowledges.

Toward the goal of obtaining such a coexistence within a Western-dominant system of education, Kanaka Maoli scholar Manulani Aluli Meyer points out that the assumed acultural and thus apolitical nature of the art of teaching and the science of learning are false. Instead, Meyer asserts that due to the marginalization of Kanaka ways of knowing as a result of colonization, Kanaka epistemology needs to be reestablished within educational systems. Emphasizing that the most important aspect of a Kanaka Maoli knowledge structure is experience, Meyer advocates an attainment of knowledge involving a form of learning not merely with the mind but also with the naʻau and through ancestral and spiritual sources (Meyer 2001). Practices, experiences, and awarenesses of place are combined into processes of knowing and researching.

Ka hālau o ke kai would follow this philosophical assertion, establishing an educational system based on a Kanaka understanding of place, history, culture, and kuleana. The mission would be to employ the knowledge within the archive of oceanic literacy, to empower local and indigenous knowledge and skills that uphold specific reflections of seascape epistemology and are based on Kanaka ontological foundations as related to ʻāina. Ka hālau would engage formal and informal student-teacher learning, emphasizing observation, listening, language, and relationship to produce an affect difficult to obtain through other educational structures.

Ka hālau o ke kai would be a gathering site where the oceanic literacy within seascape epistemology (the philosophy of knowledge based on a Kanaka way of theorizing, constructing, and validating knowledge through our historical, social, political, and spiritual relationships with the seascape) would be applied and realized. It is a place where literacy and knowledge

are theorized, created, reformulated, and encoded through seascape epistemology (Gegeo and Watson-Gegeo 2001). Ka hālau would become an organized execution of the theory within seascape epistemology, strengthening
the literacy and knowledge within the epistemology through constantly
changing and moving practices (heʻe nalu, hoʻokele, and lawaiʻa) as well as
places (ke kai).

Ka hālau o ke kai would not offer an "authentic" or "natural" connection
to or philosophy of a Kanaka way of life as related to ka ʻāina. Rather, it
would be a place where an ensemble of practices get connected, disconnected, and reconnected by Kānaka Maoli through our specific historical
context (Wood 2006). Ka hālau would become a gathering place where
Kānaka Maoli could engage oceanic literacies not only to affirm our indigenous identities but also to navigate new and future pathways through a
complex and Western-dominant world. As the coasts of Hawaiʻi encounter increased development by private and state interests, affecting not only
public access but also the ability to preserve and sustain natural resources,
ka hālau would remain a center that holds ke kai in a valued, sacred, and
familiar place. Placing ʻāina at the core of the philosophy of knowledge develops an ethics for Kānaka Maoli rooted in our ontology and epistemology,
and ka hālau could present itself as a physical center to foster this Kanaka
way of being-in-the-world.

The Hōkūleʻa

Ty Kāwika Tengan, professor of anthropology at the University of Hawaiʻi
at Mānoa, conducted research in his doctoral dissertation on a small group
of Kānaka Maoli who had formed a *hui* (group) called Hale Mua that had
been working on redefining what it means to be Kanaka. Tengan reports,
"In the case of carving, they [the Hale Mua leaders] constantly reiterate that
we are not only making a weapon but also perpetuating Hawaiian culture
and carrying on the knowledge of our kūpuna. Thus the importance lies
not in what shape the wood ends up taking but rather [in] the fact that we
go through the process and in so doing make our own mana" (Wood 2006,
47). Wood notes in relation to Tengan's work that "the practices are consti-
tutive of a living culture, not a memorial to one that is past, or a symbolic
representation of its meaning" (47–48). The oceanic literacy in hoʻokele
also performs this essential identity work for Kanaka Maoli.

The first successful voyage of the *Hōkūleʻa* from Hawaiʻi to Tahiti and back in 1976 instigated a cultural and political movement and has acted as a unifying icon for contemporary Kānaka, representing cultural resurgence and empowerment through the revival of ancestral knowledge, skills, and protocol as well as a means toward self-determination and self-identification. The waʻa kaulua has also created a powerful global connection between Kānaka and other seafaring peoples from Palau, Rarotonga, Aotearoa, Rapa Nui, Alaska, China, and Japan. This connection has allowed for an alternative means of doing politics, of creating alliances, and unifying pathways; *Hōkūleʻa* travels without linear grids or static maps, symbolically carrying messages of harmony, sustainability, and cross-cultural reciprocity and understanding.

Nainoa Thompson is realizing the vision of sailing *Hōkūleʻa* and its newly built sister escort waʻa kaulua, *Hikianalia*, around the world to eighty-five ports in twenty-six countries, and sailing over fifty thousand miles. The waʻa set sail from Hilo Bay on Saturday, May 17, 2014, to Tuatara, Tahiti, carrying with it the purpose of teaching and challenging the world's youth, particularly Hawaiʻi's youth, in the oceanic literacy of voyaging. About 30 percent of *Hōkūleʻa*'s 260 crew members are under the age of thirty. In this way, Thompson can assure that the journey and the learning experience are truly for the next generations by creating new leaders that are and will act as stewards of the planet.

Thompson explains, however, that the larger ambition of the worldwide sail is to reconnect and spread a message of peace through an indigenous vessel. He believes that by having *Hōkūleʻa* visit the world's islands, continents, and peoples, a profound message can be communicated: "There is something core about humankind that gets touched, the nerve gets touched by something like a canoe or a temple. There's something that's beyond the boundaries that we typically divide ourselves with and sometimes kill each other over. It's beyond that. It's a set of humankind values that somehow we can all embrace and all be empowered by" (Thompson 2007, 30). The Dalai Lama blessed the waʻa kaulua in April of 2012, showing his support for *Hōkūleʻa*'s mission of physically touching the world with its message of peace. A few months later, in August 2012, Nobel Peace Prize laureate archbishop Desmond Tutu from South Africa participated in a short sail on the waʻa, blessing it for the upcoming worldwide voyage. Thompson wants to spiritually understand and be with the earth to better learn and teach how to care for it. "One of our goals is just to help the world understand the importance of the ocean, and why we need to take good care of it, and how do we do that. . . . In going around the world, our main achievement is to create an extraordinary learning opportunity for thousands of youth" (Thompson 2008). Engaging global youth in such an effort, however, also offers the ability to create a new path for political, social, spiritual, and economic systems.

Within this global goal lie the internal struggles of the waʻa itself: what is the future path of *Hōkūleʻa* for the youth of Hawaiʻi? Every voyage taken by the *Hōkūleʻa* is on a continuum of a total journey toward the establishment of cultural understanding, Kanaka values, personal strength, and care for ke kai. For Thompson, the worldwide voyage is the tool to get the waʻa sailing toward these larger goals of the voyaging community. "It's an eight-year journey [including training time] of exploring, under our values that we believe in." He wants the voyage to be a profound educational experience, a novel way of transforming and evolving the current educational system in Hawaiʻi:

> The worldwide voyage, to me, is about a mechanism to create new education in Hawaiʻi because the old system is failing us in many different ways. It doesn't mean that we'd overhaul education, but it does mean that education has got to shift. So what's powerful about the worldwide voyage is that it has the gravity to bring a lot of people together. We've got the president of the university; we've got the head of the DOE; and we have the head of schools like Punahou and Kamehameha. The institutions are there, and they are, to some degree, some more than others, willing to entertain the notion of shifting. (Thompson 2008)

Thompson admits that he isn't clear on what this "shift" will look like, or even how it needs to occur, but his aspiration is to collect creative minds to discuss this concept and help formulate its realization. What is clear is the potential and compelling role of *Hōkūleʻa* in this aim. The political power of the waʻa reveals itself again; *Hōkūleʻa* is a tool for placing the "invisible" Kanaka epistemology at the forefront of Western-dominant educational and even corporate institutions. The waʻa is repartitioning the sensible through its powerful imagery and symbolism, bringing to light the marginalized Kanaka ways of knowing, learning, and being by bringing together leaders of the current power structures to think about a decentering and recentering.

Ka hālau o ke kai is a place to help initiate this recentering. It is a place to teach the specific oceanic literacy of hoʻokele to the youth of Hawaiʻi as an integrated element of education. The hope is to work with the Polynesian Voyaging Society and their vision of education through voyaging, using the Kanaka space the *Hōkūleʻa* has created, to support Kanaka epistemology. Ka hālau could incorporate the society's already existing program, Kapu Na Keiki (Hold Sacred the Children), an educational program started in 2004 to challenge and inspire students to explore and care for the ocean, coral reefs, and their islands and communities, as well as to encourage the values of compassion, giving, and service through coastal and statewide sails.

For instance, on Kapu Na Keiki's first deep-sea sail, a group of eleven teenagers ranging between the ages of sixteen and nineteen underwent two months of practice- and place-based education involving intensive navigation training. This form of oral and action-oriented learning enabled them to sail from Hanalei on July 24, 2007, to Kaʻula, a small island west of Niʻihau. As head of the program, Thompson refrained from helping the students, instead encouraging them to rely on their training and on the ocean, the heavens, and their naʻau. The students successfully found Kaʻula, and at that moment Thompson saw in the students "that wonderment of being able to use nature to guide you across the sea and the thrill of finding that destination you seek and in the process knowing you're learning about something very special, being in the wake of your ancestors. Watching them search for Kaʻula was watching them grow into themselves" (Muneno 2006, 24). Today, these Kapu Na Keiki students are among some of the leaders on the worldwide voyage, and helped to train other youth for the voyage.

Through this program it is evident how *Hōkūleʻa* acts as a working classroom, not only teaching students how to sail by traditional navigation techniques but also about identity and place in the world. While waiting for the weather to cooperate for their journey, the students on this first sail reciprocated the overwhelming embrace given them by the Hanalei community: the students helped to pull weeds in the loʻi, or taro patch, at Waipā; initiated beach cleanups; and helped to lash the Kauaʻi canoe *Nā Māhoe*. They were also required by Thompson to keep journals in which they answered the questions "Where do you want to be in twenty years?" "How will you get there?" "What makes Hawaiʻi unique?" and "Who are you?" (25). This program illustrated how practice- and place-based education involves both cognitive and embodied knowledges, both academic and experiential exercises. Ka hālau o ke kai is a permanent and physical place where, as in Kapu Na Keiki, students could learn from kūpuna about reading the stars, moon, clouds, winds, and ocean swells to physically and conceptually guide them through the seascape. It is a center where the efforts of the *Hōkūleʻa* are reinforced and perpetuated in everyday life, not as a single or segregated knowledge but as a larger epistemology.

The knowledge of hoʻokele is also digested and understood together with the knowledges of building waʻa, the protocol of cutting down a tree, stocking waʻa with foodstuffs, and the ceremonies to launch and receive the canoes and their crews, all of which were historically very spiritual processes. Ka hālau would be a place where these practices are taught, and where the value of knowing these practices is revealed. It would be a place where the knowledge within the archive can be read, practiced, and lived.

Another example is the significant knowledge of building waʻa that would be realized at the hālau. Constructing waʻa has always been a significant and core component in Kanaka culture and society, as described by Kanaka Maoli waterman and author Tommy Holmes: "Recognizing that the sea could be at once a friend and a killer, the ancient Hawaiian evolved a canoe form that may well be the most versatile and seaworthy rough water craft ever designed or built by any culture in any time—a reflection of man in supreme harmony with his environment. Of transcendent importance, the canoe in old Hawaiʻi was a nucleus, a continuum, a key to the culture" (Holmes 1981, v).

Waʻa were historically vital to Kānaka Maoli for coastal and open sea transportation as well as for harvesting the resources from the sea. According to the moʻolelo of family elders from Molokaʻi, the men who carved waʻa, kālai waʻa, were considered by most in Hawaiʻi to hold the highest mana and greatest skill. This ocean-based knowledge, however, was known by and taught to only a few, and those who did hold this literacy were considered *kāhuna*, the highest experts in their field. The term *kahuna* was not often used, only when an elderly kahuna was about to pass away would he or she designate who was to follow, breathe into that person's mouth, and pass along a bit of knowledge that had not been taught in school, thus appointing duties to the new keeper of the secret (Willis and Lee 1990). These kāhuna would search the Hawaiian koa forests to select the right tree, the one that held the spirit of the canoe. In these ways, the building of waʻa was a profoundly religious practice, involving many akua and ceremonies specifically associated with this master of artistry.

The oral tradition of passing down the skills and literacy surrounding the building and handling of waʻa through teachings and chants has largely fallen away over the years due to colonization and modernization, and much of this knowledge has been lost, although some details remain in written records and drawings by early Western explorers visiting Hawai'i, as well as by Kanaka Maoli scholars and writers. Ka hālau would offer a place where kūpuna of wa'a building would be brought to potential students, who would act as mentors and pass along their culturally critical skills. Wa'a kaulua are more than sailing vessels; they hold the lives and experiences of the crew members and builders who worked them, and this mana needs to be preserved through the literacy and spirituality of future master canoe builders.

Ka hālau would offer a place to nurture these future builders, a place to physically construct wa'a based in Kanaka customs, knowledges, and protocols while integrating contemporary materials, skills, and technologies to enhance the outcome. Modern materials were used to build *Hōkūle'a* and the newer wa'a kaulua of Hawai'i, such as *Hawai'iloa, Mo'olele, Iosepa, Makali'i,* and *Hōkūalaka'i,* mainly due to a lack of natural resources, but also for practical reasons. There are also designated escort boats with GPS devices to ensure safety on voyages, and even the methodology in teaching navigation involves Western concepts and sciences. These modernizations of voyaging allow for contemporary adaptations while embracing ancestral roots, knowledge, pride, and potential for the future. Ka hālau would emphasize a contemporary Kanaka identity by mixing modern and ancestral skills, resources, and concepts. It would bring forward the knowledge of ho'okele and provide Kānaka with the ability to physically move and travel through Oceania, independently of Western (dominant) systems. These voyaging canoes (re)teach Pacific Islanders how to integrate travel into other aspects of their lives, and what they can learn from this method of travel around, movement in, and interaction with the world.

Thompson stresses the importance of this oceanic literacy not only for cultural and personal affirmation but for environmental sustainability. The future of the oceans and this planet coincide with youth understanding, interacting with, and caring for the seascape, and much of this relationship is developed through systems of education. Thompson says,

> Up until the 1820s, the only national schools in America were for the elite, rich, and white, and it was for leadership, to run the country. In 1910, there were public schools for immigrants to be able to read, to be able to succeed in the workforce. In 1910, the whole shift was to industrialization, to industrialize this country and be the best in the world. We built the footprint to be able to do that, but at the expense of the planet, which in some ways, we are ecologically killing. So ex-

ponentially, it's going to get worse, and I don't feel the mechanisms, the politics of this country, or the economics, will stop it. So, having said all of that, what is Hawai'i's response? We did exactly what we were taught to do, industrialize, but we didn't do it with responsibility.

Hawai'i is an extraordinary school for maybe not the earth, but those who are looking for solutions. The shift in education is to look at sustainability, to look at the ecological integrity of these islands, and look at why Hawai'i is special and unique, and to protect those things. (Thompson 2008)

I propose that ka hālau become a space in which this shift in education can successfully occur. It would be a place where youth and their families can convene to learn about, build relationships with, and truly care for the seascape. In addition to cultural, individual, and spiritual affirmation, ka hālau would offer a place for people to come and learn to love the ocean. Thompson notes, "What we love, we were taught that love. If we don't understand the ocean, we won't love it. If we don't love it, we won't care for it, and if we don't care, we won't act" (Thompson 2008).

Ka hālau would require not only the implementation of the knowledge within the archive and of Kanaka place- and practice-based learning but also the certainty that a new educational system is relevant for the future. Adaptation is critical for the survival of any culture, and it must be an adaptation that balances the essence of Kānaka Maoli with the modernizing and moving world. James Clifford notes that we are reaching back to traditional knowledge and skills in voyaging and navigation and the building of canoes, and at the same time telling stories about mobility by jumbo jets. He writes, "It is an updating and reconnecting of very old traditions of inter-island contact with now something very current. Tongans, Samoans, and Hawaiians, for example, now go back and forth to Los Angeles and Las Vegas. It represents a kind of indigenous cosmopolitanism. More than simply unmaking the concept of 'the native,' it is complicating it" (Clifford 2000, 96). This cosmopolitan identification takes Kānaka Maoli beyond postcolonial theorizing that has placed the native in an imprisoned and enclosed identity that is now being broken out of, and, as a result, creating a decentered and multisited indigenous structure (Clifford 2000, 96).

The oceanic literacy of ho'okele, illustrated through the voyaging of *Hōkūle'a*, brings this discussion of decentering and multisitedness to the forefront of indigenous politics. The hope is to merge innovative and effective ideas about how to successfully "shift," recognizing not only that a shift is required but also that this shift must be relevant to the future. For instance, Thompson notes that his mentor, Mau Piailug, was chosen by his grandfather at the age of one to become a navigator. Piailug was put in tide pools in Satawal to play, "but," Thompson explains, "playing was equal to learning, to getting connected to nature" (Thompson 2008). Choosing to immerse a child in the literacy of the ocean at this age is a strategic choice when addressing the question of leadership in ocean nations in this Western-dominant reality. Yet is such an immersion relevant for the future of Hawai'i? That answer is for both the community and individuals in the Hawaiian community to discuss, but I am convinced that immersion does not have to be an exclusively Kanaka or Western act in the historical sense. There can, is, and should be a mixing.

A contemporary Kanaka epistemology and ontology includes Western elements and experiences, and a contemporary vision for Kanaka education should also include tools and knowledges from both. The critical concept in this shift is that Kanaka-based education is made as visible as, if not more visible than, a Western-based system for Kānaka Maoli. Ka hālau o ke kai may not immerse Kanaka youth in oceanic literacy at Piailug's level, but changes can begin to redirect the education in Hawai'i. Navigation is an ideal tool toward this end. Rewards from an education in oceanic literacy reach into the epistemology and ontology of future minds and hands, while simultaneously developing an ethical way of learning and approaching the world that is critical for all of modern humanity.

Nā Kama Kai

Ka hālau o ke kai would also be a place where the oceanic literacy of he'e nalu can be learned, taught, actively practiced, and philosophically engaged from a Kanaka perspective. The concept and ethics behind having a physical place to teach the act of sliding across an ocean full of waves, or to spread out and meditate, is distinct from the act of learning to surf in a Western ideology that differentiates the body from the sea. Ka hālau would specifically teach he'e nalu as an integrated practice of body, mind, spirit, and place. Kūpuna would teach students through methods of observation and experience, instructing students about the patience, time, and expression required to ride nā nalu. Students would learn that to slide across ke kai is not to control the ocean. When a wave rolls to shore, a surfer dives underneath the churning water rather than confronting it. If caught up by its power, a surfer learns not to fight the ocean and exhaust her energy and oxygen supply, but to instead relax and travel with the wave, knowing it will release her. He'e nalu requires a surfer, just as ho'okele requires a navigator, to become part of the ocean, a part of ka nalu.

Respect is expressed and taught through he'e nalu, as it is through all oceanic literacies. To ride the seascape, one must respect it. To truly respect ke kai, a surfer must first know it; she must know why the ocean moves in its seemingly arbitrary and capricious ways, and she must learn to develop her own relationship with this other living body. The student learns to hear the ocean breathe and to ride upon its inhalations and exhalations. Kanaka surfer Tom "Pōhaku" Stone puts it, "We should be learning the heartbeat of our ocean—listen to breath" (Provenzano 2007, 58). Most of the learning at ka hālau would be through this type of observation of and listening to not only kūpuna and ke kai but all of the actions and observations of past generations accumulated in the present that are manifested through language, place names, and mo'olelo about the seascape.

In 2008, professional surfer and waterman Duane DeSoto began a nonprofit called Nā Kama Kai (Children of the Sea) in an effort to offer ocean opportunities to local youth. His goal continues to be to provide access to surfboards, surf contests, and transportation to the ocean, regardless of economic or social background. Not having Kanaka blood, DeSoto was adopted by the Kanaka DeSoto family and was raised on the westside beach at Mākaha where he learned from his 'ohana, or family, about the significance and potential that the literacy within surfing could offer him and other local families in a modern world. Through his adopted family and background, he emulates Hawaiian values that are illustrated in his ambitions with Nā Kama Kai. DeSoto states, "My thing is first of all, introduce them [the youth] to the ocean. We really want to take kids that have no connection to the ocean."

He has also begun a surfboard recycling program that brings boards into the organization, fixes them up, and redistributes them to children who don't have surfboards of their own. "It can help them with boards for contests or just getting into the water. It's going to be a broken board, but it's going to be our best effort to try to introduce the ocean to the children, or help some of them get to that next level [in contest surfing]. So they can say, 'I need a longboard, I want to do a longboard contest,' and they'll have one" (DeSoto 2006).

DeSoto goes on, "Nā Kama Kai is going to be a vehicle to also see people who are meant to be in the water or to have a job associated with it. And we can have them work with that. We would like to be able to give scholarships, whether if it's to travel or go to school to be oceanographers, or help them into the lifeguard program. Whatever it is they find hard to get, we want to give them the help" (DeSoto 2006).

DeSoto is providing a resource, just as ka hālau would offer access to physical materials as well as intellectual and spiritual sources. Ultimately, he wants Nā Kama Kai to be an educational organization with an ocean awareness class in the public school system. "It's an obvious thing that's been needed for a long time, but the problem is that it's not profitable, and it's not easy" (DeSoto 2006).

The vision behind both ka hālau and Nā Kama Kai is to convince youth that they can be empowered by the ocean. DeSoto wants children to say, "I can be a person filled with confidence that says I can be a governor, I can be a senator, I can be a lifeguard, I can be a fireman." That's the empowerment that comes with being a surfer, DeSoto says. "You're out there on your own, and you're negotiating the ocean, and you're out there depending upon yourself for survival. It creates a lot of strength and courage in you." No matter one's surfing ability, a surfer in the ocean is always self-sufficient and self-reliant. The oceanic literacy of he'e nalu truly teaches an individual how to find the knowledge and strength within oneself, guided by cultural and ancestral skills, to glide or "surf" autonomously through a globalized world.

Teaching he'e nalu in ka hālau, as with the literacy of ho'okele, would begin with an awareness of, respect for, and fear of the ocean. DeSoto says, "I encourage fear; I encourage them to be afraid. Not afraid to hide and crawl in a shell, but to be afraid and then to tame their fear, because nobody in the world who is successful and goes big is not afraid. . . . It's how you tame and control your fear. We teach them to feel good about being afraid; that means you actually respect her [points to the ocean], you care" (DeSoto 2006).

Once a student has learned a respectful fear of ke kai by engaging and interacting with it, all of the details within the literacy of he'e nalu are revealed. She can feel currents, smell the salt and limu, hear the waves, see the reef underneath her board, and read the movement of the foam on the water's surface. She can understand the physics behind standing up on a moving wall of water, and can balance and steer with the push and pull of the wave. He'e nalu, like all oceanic literacies, is experienced and absorbed, just as cultural knowledge is learned, through observation and immersion.

Ancestors, both present and past, show themselves as guides to students in their development of skills and protocol, and with the rift in this link due to colonization, revitalizing this form of cultural learning is critical if we, as Kānaka, are to truly understand ourselves. Our history is born from ke kai. The current lack of confidence in Hawaiian culture necessitates places and times such as ka hālau to allow for a return to a relationship with, knowledge of, and caring for the seascape.

The critical and central role of the kūpuna is revealed in ka hālau, where their guidance and mentorship would do more than simply pass down information and knowledge. Kūpuna would also pass down their mana to students through example and as natural leaders. DeSoto explains the process of passing down and learning an oceanic literacy through Kanaka epistemology:

> You have to live something to even try to communicate the right message to someone. You can't tell someone how to sail and you're not that good at it. It's like these surf guys [points to commercial surf instructors in Waikīkī], they're trying to teach someone how to surf, and they don't even know how to surf. Their students are never going to get the whole essence of it. And when I do my lessons, my clinics, I bring down professional water people because, automatically, the mana is translated through them just standing there. And then everything else is just learning the details now, because they're automatically getting that energy, that confidence in the water, and that's having a basis, something legitimate and real. . . . You have so many guys so confused with their own culture that they just become capitalist and it shows. . . . They're missing the point. (DeSoto 2006)

Ka hālau would support and supplement the goals of Nā Kama Kai by re-inscribing an indigenous place with Kanaka knowledge and values. Again, it is a decentering and a recentering.

Loko I'a

In addition to experiential learning, moʻolelo and storytelling would also be fundamentally incorporated into educational applications in ka hālau o ke kai. Students would listen to kūpuna narrate histories that sit within ahupuaʻa to teach students where *moi* (Pacific threadfin) holes are located, and how to care for and respect moi and their habitat. Ka hālau would approach learning as a connection between knowledge and responsibility, demonstrated through the oceanic literacy of lawaiʻa.

The oceanic literacy of lawaiʻa embraces specific cultural protocols and fishing techniques, and the building and paddling of waʻa, and it also provides a major source of protein for Kanaka Maoli both historically and today. It is a literacy that also teaches confidence as related to identity and place, as well as self-reliance. Lawaiʻa is a powerful literacy integrated into issues of sustenance and survival for Kānaka, and in a contemporary world where fish populations are decreasing, pollution and carbon dioxide are killing reefs, and climate change is affecting the survival of ecosystems, sustainable knowledge is not only culturally empowering, it is a necessity.

Through great care and patience, Kanaka ancestors have identified historic fishing grounds. The descriptions of the zones and moods of the sea recognized by Kānaka Maoli are almost as detailed as their terrestrial equivalents (D'Arcy 2006). Ka hālau would teach students to recognize, for instance, the various sections of coral reefs. Students would learn the historical zones, which were named according to their color and motion, the wave action in the area, tidal patterns, and the type of fishing conducted in the area: "Coral reefs were divided into fishing grounds such as papa heʻe (octopus grounds) and kai ʻōhua (feeding grounds of young fish). Waters farther from the shore were largely distinguished as fishing grounds. Thus, kai paepae were sea areas for pole fishing, and kai lūheʻe, squid fishing areas. These seas were known generally as kai uli (blue seas)" (46).

Kanaka ancestors' knowledge of fishing grounds was exceptional. The writings of A. D. Kahaulelio in the Hawaiian newspaper *Nupepa Kuokoa* in 1902 reflects Hawaiian knowledge of fish and sea floor characteristics as deep as two hundred fathoms. Kahaulelio knew a hundred fishing grounds that were ten or more fathoms deep. The following is a description of part of the seabed between the islands of Maui, Lānaʻi, and Kahoʻolawe:

The kinds of fishing grounds both deep and shallow are as follows: From Point Hawea at Kaanapali to Lae-hima-lani point, the fishing grounds are very shallow, from twenty to thirty fathoms in depth. It is also true with these that at close to the writers dwelling place in front of Kamaiki point, the depth is the same. In between these places the sea floor is flat with no cliffs and mountains that are overgrown with trees that grow in the sea. From Launiupoko to Papawai Point the sea outside of them hold most of the fishing grounds and contain some deep depressions good for *kaka* [long line] fishing. Allowing six feet to a fathom, they are 1,200 feet deep and that is about the height of a mountain in the ocean all grown over by *ekaha* trees. With these mountains in the sea, the lines and hook often get entangled among the trees. They have many branches and leaves and find a sale among sea captains because they think it strange that trees grow in the ocean. The fishing ground called Laepaki (Kealaikahiki) is five miles distant, from fifteen to twenty fathoms deep, that is the shallowest one. It is only fifteen fathoms deep. The sea floor and the fish swimming to and fro are plainly visible and that is one of the most productive of the three fishing grounds of Kahoolawe. (Kahaulelio 1902, 24–25)

The oceanic literacy of fishing includes knowledge of the location of fresh water springs flowing into the ocean, how to paddle canoes noiselessly when fishing for *ulua*, or giant jack trevally, preparing fishing canoes with offerings at fishing shrines, and the use of *kū'ula* (a stone god used to attract fish) before and after an expedition to appease the fishing god Kū'ula Kai as well as personal 'aumākua, or ancestral gods. Knowledge of fishing kapu remains particularly critical for Kānaka Maoli today not only for ceremonial and spiritual purposes but also for ecological sustainability.

In the ecological cycle, most reef fish spawn when currents sweep their eggs out to sea, where the young fish hatch, band into schools, and then move back in to shore. In tow are the bigger fish, such as ulua and moi, and this season is very prosperous for shore fishing. Familiar with these cycles, Kānaka Maoli established a cultural system in tandem with the ecological one in which fishing was kapu during those observed times of the year when fish were spawning. Angela Hi'ilei Kawelo, director of Paepae o He'eia, He'eia fishpond in Kāne'ohe, O'ahu, says of the fishing kapu system in Hawai'i: "The konohiki [the headman of an ahupua'a under the ali'i] would assess the resources and what they're doing. That was his kuleana. Based upon what he saw out there, understanding that every year is different . . . and that the spawning seasons change, they might be off by a month or so. But the konohiki would go out and assess the resources and then place kapu depending upon what those resources were doing. The general kapus were during the times of spawning, to assure there's always keiki" (Kawelo 2009). Management of these kapu, which differed in each ahupua'a on each island, were the kuleana of the konohiki to determine when fish populations needed to be conserved. It was the responsibility of the konohiki to assure the people of Hawai'i that food sources would not diminish.

Ka hālau o ke kai would teach students not only the spawning times of different fish, the kapu (prohibition), but also how to recognize when fish are spawning. Students will also learn to read sand movements and how such changes affect fish behavior. As large quantities of sand move during different seasons of the year due to current and swell action, as seen at Pūpūkea Reef on O'ahu and Mo'omomi coast on Moloka'i, marine species also shift locations as their habitats fill with sand. This change in habitat can also trigger fish spawning activity, and students would be able to read all of this movement through experiential and place-based education of oceanic literacy.

Conservation through spawning seasons has always been an exceptionally critical aspect of the oceanic literacy of lawai'a for Kānaka Maoli. Due to the extreme isolation of the Hawaiian Islands, marine life there is less diverse than near other Pacific Islands, and there is a lack of nutrients due to a less extensive reef habitat. As a result, Hawai'i has had less favorable ocean currents and upwellings for fish populations (Holmes 1981). The production and maintenance of fishponds as aquaculture was thus vital to sustain the Hawaiian population in the early 1800s, which was estimated to be between 175,000 and 225,000 (109). *Loko i'a* (fishponds) were another historical way that Kānaka Maoli helped to harvest and replenish fish, and they were reliable resources used to deal with the uncertainties of the marine environment.

Loko i'a were generally stone constructions along shallow coastlines. The main fish kept in these ponds were the herbivorous fish *awa* (milkfish) and *'ama'ama* (mullet) because herbivorous feed on the naturally growing plants abundant in the ponds rather than other fish, which was a much more effective use of their marine sources. A study suggested that these fishponds yielded an average of 166 kilograms of fish per acre. He'eia fishpond has a water space of 88 acres, with the stone wall, built in a full circle, stretching 1.3 miles, or 7,500 feet, with a double-sided rock with coral fill (Kawelo 2009). Kawelo explains that fishponds such as He'eia made in the *kuapā* (shore ponds) styles, encloses the natural estuary and habitat in a very efficient and naturally productive system:

If you can imagine what our kūpuna were seeing, were these areas of brackish water, highly productive environments, with a lot of primary producers, meaning algae and phytoplankton. Noticing that a lot of the *pua* [young] fish survive in these kinds of environment, the action of closing in an environment like that, you're able to enclose it, but you're also able to enhance it. You're able to regulate what goes in and out. You're able to keep fish in, but to also recruit fish into the fishpond. That's the primary function, to grow fish.

As such, loko iʻa were a highly concentrated and reliable resource. Kanekoa Kukea Schultz, Kāneʻohe Bay coordinator for The Nature Conservancy, notes that estuaries are the most productive systems in the world. In an estuary, fresh water from rivers and streams flow into the ocean, mixing with the salt water and creating a habitat hospitable to numerous different species. Schultz says, "In one environment, there are certain species that maximize and out-compete the others, so with small fluctuations in an estuary, wider ranges of habitat increase" (Schultz 2009).

The potential site for ka hālau at Kaloko Beach along the Ka Iwi coast is a historical fishpond site. Ka hālau o ke kai could help to rebuild these ponds using one of the six traditional styles of stacking, which would be determined by the resources currently available in the area as well as the water conditions. The benefits are truly endless for students as well as ka ʻāina. For instance, students will learn about physics through the construction of sea walls, tilting them to let the waves roll up and over them. Kawelo says, "I think for the local kids, science has always been a scary subject. I don't know why, because it's inherent in us; our kūpuna were scientists, they did things through observation. For the keiki to see that our kūpuna designed something like this [Heʻeia fishpond] based off of observation, and that they understood how the environment affects the habitat and the ecosystem, and knowing the species that exist in the pond at an intimate level, then they're able to see that science doesn't have to be scary" (Kawelo 2009).

Students also learn the critical element of mālama ʻāina, of respecting and caring for the islands rather than governing them. Kawelo explains that loko iʻa involve a "hands-off" approach to cultivating fish. The environment is facilitated, but it is not regulated. Once the pond is built, the fish take care of themselves with the natural resources available in the area. She notes, "There's really no input that is needed other than to ensure that fresh water is making its way to the ocean" (Kawelo 2009). Kānaka Maoli are not controlling the fish; they simply create a favorable environment in which the fish can grow.

In addition to rebuilding and overseeing the fishponds, students at ka hālau learn to identify and catch fish, as well as clean, share, and even eat fish so that nothing is wasted and the animals are honored for their sacrifice. These practices exemplify respect and caring for ka ʻāina, as well as the concept of kuleana, all of which are critical not only for Kanaka identity and self-determination but also for environmental and ethical purposes. Imaikalani Kalahele, a Kanaka poet, artist, and fisherman, said,

> Even in my time I remember seeing big schools of fish. Talking to the old guys when they followed the mullet down on the windward side of the island [of Oʻahu], they could see these big clouds of fish just moving. . . . That doesn't happen anymore. That was in my time.
>
> I took my grandchildren to the beach the other day. It used to be fertile ground for squid and good *limu* (seaweed) was plentiful. Now it's all this weird rubbish that came from some place else and it covers the entire reef. This is industrial pollution. I think if we had a different understanding of what our relationship really is—not just *ʻāina*, I'm talking about *Papa* (the Earth)—it's a different understanding of how we relate to *Papa*. For me, the Pacific is our back yard. Sometimes when you use the term ocean, it tends to turn it into this other thing, yet it really is our back yard. We don't have fences between us and our cousins in Samoa, we just have really long back yards. (Provenzano 2007, 114)

Ka hālau would foster this relationship to Papa and Wākea by teaching students to act as stewards of the reef, to help maintain fish populations, and to help eradicate invasive limu species that smother reefs and drive out native fish populations. Ka hālau would be a place that incorporates the skills associated with this connection to place: the carving of fishhooks and education about the different sizes and shapes of the hooks as related to the different fish species. Not all of the hooks are made from traditional materials for practical matters and environmental consideration; most are made from cow bone and shells, and some are wood. Kanaka artist and craftsman, Gordon ʻUmialiloalahanauokalakaua Kai, says, "I teach making traditional fish hooks—making with modern tools but maintaining the integrity of the traditional design—fish traps and other fishing techniques" (99). Hooks were historically made from pearl shell, turtle shell, human or animal bone, and sometimes wood, as used in large composite hooks for shark fishing (Hiroa 1957).

Lashing hooks and waʻa for fishing, sewing *lau* (nets), spearing octopus, diving for *weke ʻula* (orange goatfish), and learning to read the textures of the ocean for fish activity are other Kanaka skills learned and practiced at ka hālau. Yet, at ka hālau, these skills are all practiced in tandem with the use of motor boats, metal or plastic lures, and GPS devices, allowing students to extract from both historical and contemporary methods and materials to determine what they believe is most beneficial for them individually in their diverse futures and purposes. Although techniques may evolve and change, the epistemology of oceanic literacy remains one of connecting to and creating an intimate relationship with the seascape. The ability to feed oneself is at the core of self-determination, and while all Pacific Islanders cannot necessarily become self-sufficient through lawaiʻa in today's world, simply obtaining the knowledge of how to do so is empowering. Applying the theory of seascape epistemology in a place such as ka hālau o ke kai so that Kānaka Maoli can move through a modern and globalized world provides them with empowering tools. Simply interacting with the ocean, listening, watching, smelling, and touching ke kai enables not only the literacy but also the philosophy of knowledge that is rooted yet flexible; that is based in years of accumulated experience, data, and awareness; and that encourages harmony rather than difference and dominance.

Theory Inside the Seascape

The seascape is a living classroom with kūpuna, fish, birds, wind, clouds, rain, the moon, and ocean waves serving as teachers. Students need all of these kumu, or teachers, because kūpuna alone can't instruct students in how to understand, read, or connect to ke kai in all of its colors and changing forms, just as the birds and fishes alone cannot provide sufficient observation of the different times and spaces within the sea. Kānaka Maoli are encouraged to embrace the entire seascape through oceanic literacy and the philosophy of seascape epistemology, completely connecting with and drawing upon the political, social, economic, recreational, spiritual, and cultural endowment and potential of the ocean.

The ambition of ka hālau o ke kai is to assert Kanaka identity and culture through a Kanaka epistemology. Application of epistemology is critical for Kānaka Maoli because it offers a foundation for knowledge as a recognized theory or justified belief. Theory helps to explain indigenous existence in contemporary society, and, as professor of indigenous education at The University of Waikato in New Zealand, Linda Tuhiwai Smith argues that because indigenous theory is grounded in "a real sense of, and sensitivity towards, what it means to be an indigenous person," theory can help indigenous peoples to determine priorities, strategize, and take control of their resistances (Smith 1999). Theory allows indigenous peoples to design their own tools to achieve their own goals. The theoretical significance of epistemology also organizes knowledge into an ontology, a way of existing. Progress in other areas of philosophy often depend upon epistemological presuppositions, so to offer seascape epistemology based in indigenous place broadens philosophical ways of knowing and being through an indigenous frame.

When applied by ka hālau, seascape epistemology could reach beyond academia, while using the theories within it, and carry it into the community of Hawai'i. Seascape epistemology could engage and care for community spaces, community members, and values while embracing historical forms of education. Yet articulating, documenting, and analyzing Kanaka culture and literacies is not just about retrieving something of the past; it is also about teaching contemporary students how to observe and act creatively and ethically. The cultural hope of ka hālau is to create sensitive, well-rounded, moral, and interested individuals. The political ambition is accepting, relearning, and honoring indigenous and alternative ways of interacting with the world. Thompson's response to ka hālau is optimistic:

> If the old system of education took us to a place that is not healthy for the earth [or Kānaka Maoli], it's the perception that it needs to shift. We need a new school. So to have a school about the ocean for young children, that makes pure sense. Why would we not do it? If the oceans die, we die. From the lens of looking at what a new school should look like to make sure the shift is relevant in adapting to the time, why wouldn't we have an ocean school for young kids? (Thompson 2008)

The answer, of course, is that the current political and social structures don't endorse this shift because it decenters the locus of power. A hālau, locally established in Hawai'i, by Kānaka Maoli and for Kānaka Maoli, is not a solution, it is an engagement. It is a critical step in the voyage toward empowerment and identity affirmation through the application and realization of seascape epistemology. The purpose behind the imagination of seascape epistemology is to transform knowledge and being—not merely to educate but to spiritually and intellectually transform. Thompson knows this journey of transformation is possible: "You know how I know, because it did it to me."

Humanity is found in the sea. "Humanity": human beings collectively; the fact or condition of being human; humaneness. But if we are born from the ocean, and Kānaka Maoli ancestors are still living in its depth as coral polyp and sandpaper sharks, is the seascape also human? It is at least as complex and alive, and it affects our souls, creating states of wonder that we call being "human." The sea is also a body that can be intuited only through the poetic:

Waka 93

My face is broken by the waves.
I am the sea, ocean, giver and taker,
primordial pre-culture pre-life.

To define me is to limit me,
one may as well define the planet.

Yet I am delicate, can feel a piece of wood
slip across my eye, can feel the calls
of men rowing as they dip into me

as if I was a well to scoop from.
Some of these I have taken
into the waters of my being.

So I am part human.
(Sullivan 1999, 103)

Finding the words to express the seascape as it wets my skeleton and salts my veins is a thirst that drives me. Interacting with this swirling life form taps me into unseen possibilities. Attempting to articulate our relationships with nature, with the ocean, is to be human. That is why humanity is found in the sea. I am the moonlight that shines from the black heaven, dispersed through the watery prism of swells into another realm. The unseen can be seen in my imagination as a being both integrated and free. I can become my own process of becoming within this universe unto itself, with life, rhythms, colors, and sounds unique to this watery sphere. Inward I go.

Notes

Introduction

1 Aware of the contradiction indigenous studies faces when speaking about the Native while struggling to resist essentialism, this work aims to contribute to the growing effort to demarcate alternative and multisited spaces in which indigenous peoples can construct autonomous identities.

2 Kēhaulani Kauanui notes in her book, *Hawaiian Blood: Colonialism and the Politics of Sovereignty and Indigeneity*, "There is not one accepted founding cosmological narrative of the Hawaiian world. The Kumulipo is a prominent genealogy of the universe that came to rule the Hawaiian origin genealogies, but there are a number of other possibilities to choose from" (Kauanui 2008, 23).

3 The reference to "roots and routes" was first used by Paul Gilroy in his work *The Black Atlantic: Modernity and Double Consciousness* (1993), as well as James Clifford in his book *Routes: Travel and Translation in the Late Twentieth Century* (1997).

4 *Ka'ao* is the term for fictional stories, and *mo'olelo* is the term for a narrative about a historical figure, one that is supposed to follow historical events. Martha Beckwith explains, "Stories of the gods are moolelo. They are distinguished from secular narrative not by name, but by the manner of telling. . . . Folktale in the form of anecdote, local legend, or family story is also classed under moolelo" (Beckwith 1970, 1). The distinction between *ka'ao* and *mo'olelo*, however, should not be too literal; the distinction is in the intention of the narration rather than in the facts.

Chapter 1. HE'E NALU

1 *Honolulu Advertiser*, November 27, 2005.

2 According to the Department of Business, Economic Development and Tourism, in 2012, the state of Hawai'i welcomed 615,675 tourists in May (397,430 of whom came to O'ahu), and 677,218 in June (425,482 to O'ahu).

3 Again, my aim is not to idealize or demonize individuals within cultures. Instead, I'm identifying and illuminating the fact that there remain thought-worlds that continue to dominate and subjugate nations, cultures, and individuals. These thought-worlds include colonial and economic colonizations, capitalism, militarism, and identity politics. Samantha and Kula are emblematic of these larger systems and philosophies rooted in specific historical events and institutions.

4 The Moana Pier was demolished in 1930 due to deterioration.

5 The imagery in this quote speaks to Teresia Teaiwa's article "bikinis and others/pacific and n/oceans." In this piece she writes, "Bikini Islanders testify to the continuing history of colonialism and ecological racism in the Pacific basin. The bikini bathing suit is testament to the recurring tourist trivialization of Pacific Islanders' experience and existence. . . . The bikini bathing suit manifests both a celebration and a forgetting of the nuclear history of Pacific Islanders" (Teaiwa 1994, 87). The surf tourism industry is also radioactive, both celebrating and forgetting (through images of the bikini, as well as the surfboard, white sand, and consumable waves) the colonial history of Pacific Islanders.

6 The *Endless Summer* movies widely invoked the idea of colonization in surfers' minds by establishing the "ideal surf lifestyle" as one that involved travel around the globe in search of the "endless summer," an endless holiday.

7 Translated from the newspaper *Ka Leo o ka Lahui*, March 24, 1893.

8 From the Hawaiian newspaper *Ke Aloha ʻĀina*, August 22, 1919.

9 See Isaiah K. Walker's "Hui Nalu, Beachboys, and the Surfing Boarder-Lands of Hawaiʻi."

Chapter 2. OCEANIC LITERACY

1 In searching for a Kanaka author of this moʻolelo, I found that those Kanaka sources that do retell the story of Māmala or talk about the area, ke kai o Māmala, which was named after this *kupua* (demigod), all cite William D. Westervelt as their source: Pukui and others in *Place Names of Hawaii*; George S. Kanahele in *Waikīkī, 100 B.C. to 1900 A.D.: An Untold Story*, Honolulu: The Queen Emma Foundation (1995); and the Bishop Museum Hawaiian Ethnographic Notes collection under "ke-kai-o-Mamala, Kona, Oahu."

2 For a more detailed history on the beginnings of Hōkūleʻa, read Ben Finney's *Voyage of Rediscovery* (1994) or visit the Polynesian Voyaging Society's website at http://pvs-hawaii.com/.

Chapter 3. SEASCAPE EPISTEMOLOGY

1 It should be clarified that makaʻāinana could have mana independent of social rank. For instance, a skilled fisherman would be said to have mana, as is seen through the moʻolelo of Mākālei. However, if an aliʻi was not pono, he or she could be killed, as illustrated in the moʻolelo of Kaʻū kū Mākaha. Mana can be gained throughout one's life by excelling in an area, such as heʻe nalu, navigation, or war, all of which

required skills and knowledges recognized and coveted by both the community and the gods.

2   The concept of "self" is expressed here not as an individual or isolated concept of existence in itself but as an existence with regard to knowing the world, or a modality of being-in-the-world. I include a brief discussion of diverse constructions of the self here because I believe it is constructive in my development of a Kanaka way of knowing and existing in relation to ke kai. It should be noted, however, that seascape epistemology moves beyond a constructed identity or singular concept of self into an existence that is constantly interconnected with rhythms and flows of places as metaphorically illustrated through the seascape.

3   A complete discussion on the Kanaka concept and arrangement of time is provided by Rubellite Kawena Johnson in *Kumulipo: The Hawaiian Hymn of Creation*.

Chapter 4. HOʻOKELE

1   Thompson's star compass organizes the rising and setting points of stars. The concept is that stars rise in the east, arc overhead, and set in the west, defining points on the horizon to steer by. While looking due north or south, the stars appear to circle around the north and south celestial poles. Their rising and setting define additional points, which Thompson (like Piailug) divided into thirty-two equal arcs, which he calls "houses," similar to a mariner's magnetic compass. North is marked by Hōkūpaʻa (Polaris), which is nearly motionless, and south is defined by pointer stars such as Gacrux and Acrux in the Southern Cross, which point to the south celestial pole. East and west are marked off 90 degrees from these two coordinates. The quadrants are further divided into points 11 degrees and 15 minutes apart to define the "houses" of his compass (Howe 2006, 188–89). For more on Thompson's star compass, see Howe 2006.

2   Jean-Baptiste de Lamarck was a nineteenth-century French evolutionist who is associated with a theory of heredity or the inheritance of acquired traits. A Lamarckian process asserts that cultural evolution proceeds through the passage of acquired attributes.

Chapter 5. KA HĀLAU O KE KAI

1   As the focus of this work is on one means of affirming political identity for Kānaka Maoli, I present ka hālau o ke kai specifically for them. I imagine ka hālau, however, will be open to and beneficial for everyone who has made Hawaiʻi their home.

2   Although focus is not placed on the oceanic literacy of lawaiʻa in this work, I do include it briefly in this discussion because it is such a critical ocean-based knowledge, and one which I believe should be part of an ocean education center. In-depth research has been and continues to be done on the significance of lawaiʻa regarding Kanaka epistemology and ontology.

Alcoff, Linda Martin. 2005. "Foucault's Philosophy of Science: Structures of Truth/ Structures of Power." In *Blackwell Companion to Continental Philosophies of Science*, edited by Gary Gutting, 211–23. Oxford: Blackwell.

Alliez, Éric. 1996. *Capital Times: Tales from the Conquest of Time*. Translated by Georges Van Den Abbeele. Minneapolis: University of Minnesota Press.

Anthony, Naʻalehu. 2008. Interview by author, February 22, tape recording, Honolulu, Hawaiʻi.

Atkins, Kim. 2003. "Paul Ricoeur (1913–2005)." *Internet Encyclopedia of Philosophy*. Accessed September 12, 2014. http://www.iep.utm.edu/ricoeur/.

Bacchilega, Christina. 2007. *Legendary Hawaiʻi and the Politics of Place: Tradition, Translation, and Tourism*. Philadelphia: University of Pennsylvania Press.

Barilotti, Steve. 2002. "Lost Horizons: Surf Colonialism in the Twenty-First Century." *Surfer's Journal* 11 (3): 89–97.

———. 2005. "Localism Works." *Surfermag.com* 44 (5). Accessed December 3, 2005. http://www.surfermag.com.

Basso, Keith. 1996. *Wisdom Sits in Places: Landscape and Language among the Western Apache*. Albuquerque: University of New Mexico Press.

Beckwith, Martha W. 1970. *Hawaiian Mythology*. Honolulu: University of Hawaiʻi Press.

Benítez-Rojo, Antonio. 1996. *The Repeating Island: The Caribbean and Postmodern Perspective*. Durham, NC: Duke University Press.

Bennett, Jane. 1994. *Thoreau's Nature: Ethics, Politics, and the Wild*. Thousand Oaks, CA: Sage Publications.

Berlin, Isaiah. 1998. *The Proper Study of Mankind: An Anthology of Essays*. Edited by Henry Hardy and Roger Hausheer. New York: Farrar, Straus, and Giroux.

Blankenfeld, Bruce M. 2008. Interview by author, March 20, tape recording, Honolulu, Hawaiʻi.

Brown, Bruce, dir. 1964. *The Endless Summer.* Bruce Brown Films, Torrance, CA.

Brown, DeSoto. 2006. *Surfing: Historic Images from Bishop Museum Archives.* Honolulu, HI: Bishop Museum Press.

Buckley, Ralph. 2002a. "Surf Tourism and Sustainable Development in Indo-Pacific Islands. I. The Industry and the Islands." *Journal of Sustainable Tourism* 10 (5): 405–24.

———. 2002b. "Surf Tourism and Sustainable Development in Indo-Pacific Islands. II. Recreational Capacity Management and Case Study." *Journal of Sustainable Tourism* 10 (5): 425–42.

Carson, Rachel L. 1989. *The Sea around Us.* New York: Oxford University Press.

Carter, Paul. 2009. *Dark Writing: Geography, Performance, Design.* Honolulu: University of Hawai'i Press.

Charlot, John. 2005. *Classical Hawaiian Education: Generations of Hawaiian Culture.* La'ie: Pacific Institute, Brigham Young University–Hawaii.

Clark, John R. Kukeakalani. 1977. *The Beaches of O'ahu.* Honolulu: University of Hawai'i Press.

———. 1990. *Beaches of Kaua'i and Ni'ihau.* Honolulu: University of Hawai'i Press.

———. 2007. Interview by author, January 26, tape recording, Honolulu, Hawai'i.

Clifford, James. 1989. "Notes on Travel and Theory." *Inscriptions* 5. Accessed October 8, 2015. http://ccs.ihr.ucsc.edu/inscriptions/volume-5/.

———. 2000. "Valuing the Pacific" (interview-essay). In *Remembrance of Pacific Pasts*, edited by Robert Borofsky, 92–101. Honolulu: University of Hawai'i Press.

———. 2001. "Indigenous Articulations." *Contemporary Pacific* 13 (2): 468–90.

Coleman, Stuart Holmes. 2001. *Eddie Would Go: The Story of Eddie Aikau, Hawaiian Hero.* Honolulu, HI: Mind Raising Press.

Corbin, Alain. 1994. *The Lure of the Sea: The Discovery of the Seaside in the Western World, 1750–1840.* Berkeley: University of California Press.

Dancy, Jonathan, and Ernest Sosa, eds. 1992. *Blackwell Companions to Philosophy: A Companion to Epistemology.* Oxford: Blackwell.

D'Arcy, Paul. 2006. *People of the Sea: Environment, Identity, and History in Oceania.* Honolulu: University of Hawai'i Press.

DeLanda, Manuel. 2006. *A New Philosophy of Society: Assemblage Theory and Social Complexity.* London: Continuum Books.

Deleuze, Gilles. 1989. *Cinema 2: The Time Image.* Translated by Hugh Tomlinson and Robert Galeta. Minneapolis: University of Minnesota Press.

Deleuze, Gilles, and Felix Guattari. 1987. *A Thousand Plateaus: Capitalism and Schizophrenia.* Translated by Brian Massumi. Minneapolis: University of Minnesota Press.

Desmond, Jane C. 1999. "Picturing Hawaii: The Ideal Nature and the Origins of Tourism, 1880–1915." *Positions: East Asia Cultures and Critique* 7 (2): 459–502.

DeSoto, Duane. 2008. Interview by author, October 24, tape recording, Honolulu, Hawai'i.

Dhareshwar, Vivek. 1989. "Toward a New Narrative Epistemology of the Postcolonial Predicament." *Inscriptions* 5. Accessed October 8, 2015. http://ccs.ihr.ucsc.edu.

Diaz, Vincent M. 2008. Interview by author, June 3, email.

————. 2009. "Moving Islands of Sovereignty." Forthcoming in *Sovereign Acts*, edited by Frances Muntaner-Negron. Cambridge, MA: South End Press.

Diaz, Vicente M., and J. Kēhaulani Kauanui. 2001. "Native Pacific Cultural Studies on the Edge." *Contemporary Pacific* 13: 315–41.

Dougherty, Carol. 2001. *The Raft of Odysseus: The Ethnographic Imagination of Homer's Odyssey*. Oxford: Oxford University Press.

Dreyfus, Hubert J. 1991. *Being-in-the-World: A Commentary on Heidegger's Being and Time, Division 1*. Cambridge, MA: MIT Press.

Farber, Thomas. 1994. *On Water*. Hopewell, New Jersey: Ecco Press.

Fermantez, Kali. 2007. "Between the Hui and Da Hui Inc.: Incorporating N-Oceans of Native Hawaiian Resistance in Oceanic Cultural Studies." *Indigenous Encounters: Reflections on Relations between People in the Pacific* 43, 85–99.

Fernandes, Jorge Luis Andrade. 2002. "The Return of the Native: Postcolonial Migrancy and the (Im)Possibility of the Nation." PhD diss., University of Hawai'i at Mānoa.

Fernandez, Ramona. 2001. *Imagining Literacy: Rhizomes of Knowledge in American Culture and Literature*. Austin: University of Texas Press.

Finney, Ben, and James D. Houston. 1996. *Surfing: A History of the Ancient Hawaiian Sport*. San Francisco: Pomegranate Artbooks.

Fluker, Martin. 2002. "Riding the Wave: Defining Surf Tourism." Submitted to the CAUTHE conference, Coffs Harbour, Australia, August 2002.

Gegeo, David. 2001. "Cultural Rupture and Indigeneity: The Challenge of (Re)Visioning Place in the Pacific." *Contemporary Pacific Journal* 13 (2): 491–507.

Gegeo, David Welchman, and Karen Ann Watson-Gegeo. 2001. "'How We Know': Kwara'ae Rural Villagers Doing Indigenous Epistemology." *Contemporary Pacific* 13 (1): 55–88.

Gladwin, Thomas. 1970. *East Is a Big Bird: Navigation and Logic on Puluwat Atoll*. Cambridge, MA: Harvard University Press.

Halualani, Rona Tamiko. 2002. *In the Name of Hawaiians: Native Identities and Cultural Politics*. Minneapolis: University of Minnesota Press.

Haraway, Donna. "The Privilege of Partial Perspective." *Feminist Studies* 14 (3): 575–601.

Hau'ofa, Epeli. 1993. "Our Sea of Islands." In *A New Oceania: Rediscovering Our Sea of Islands*, edited by Epeli Hau'ofa, Eric Waddell, and Vijay Naidu, 2–16. Suva, Fiji: School of Social and Economic Development, University of South Pacific, in association with Beake House. Reprinted in *Contemporary Pacific* 6:147–61.

————. 2000. "Epilogue: Pasts to Remember." In *Remembrance of Pacific Pasts: An Invitation to Remake History*, edited by Robert Borofsky, 453–71. Honolulu: University of Hawai'i Press.

————. 2005. "The Ocean in Us." In *Culture and Sustainable Development in the Pacific*, edited by Anthony Hooper, 32–43. Canberra: ANU E Press.

————. 2008. *We Are The Ocean: Selected Works*. Honolulu: University of Hawai'i Press.

Heidegger, Martin. 1962. *Being and Time*. Translated by John Macquarrie and Edward Robinson. San Francisco: Harper and Row.

Hereniko, Vilsoni. 1999. "Representations of Cultural Identities." In *Inside Out: Literature, Culture Politics, and Identity in the New Pacific,* edited by Vilsoni Hereniko and Rob Wilson, 137–66. Lanham, MD: Rowman and Littlefield.

———. 2000. "Indigenous Knowledge and Academic Imperialism." In *Remembrance of Pacific Pasts: An Invitation to Remake History,* edited by Robert Borofsky, 78–91. Honolulu: University of Hawai'i Press.

Hiroa, Te Rangi (Peter H. Buck). 1957. *Arts and Crafts of Hawaii.* Honolulu, HI: Bishop Museum Press.

Holmes, Tommy. 1981. *The Hawaiian Canoe.* Hanalei, HI: Editions Limited.

Holt-Takamine, Vicky. 2007. Interview by author, January 29, tape recording, Honolulu, Hawai'i.

Ho'oulumāhiehie. 2006. *Ka Mo'olelo o Hi'iakakapoliopele: The Epic Tale of Hi'iakakapoliopele.* Translated by Marvin Puakea Nogelmeier. Honolulu, HI: Awaiaulu Press.

Howe, K. R., ed. 2006. *Vaka Moana: Voyages of the Ancestors.* Honolulu, HI: University of Hawai'i Press.

Hui Mālama o Mo'omomi. 2008. "Pono Fishing Calendar." Funded by U.S. Administration for Native Americans. Mo'omomi, HI: Moloka'i.

Iaukea, Sydney. 2008. Interview by author, September 17, email.

Johannes, R. E. 1981. *Words of the Lagoon: Fishing and Marine Lore in the Palau District of Micronesia.* Berkeley: University of California Press.

Johnson, Rubellite Kawena. 1981. *Kumulipo: The Hawaiian Hymn of Creation.* Honolulu, HI: Topgallant.

Jolly, Margaret. 2001. "On the Edge? Deserts, Oceans, Islands." *Contemporary Pacific* 13 (2): 417–66.

Kahaulelio, A. D. 1902. "Fishing Lore." *Ka Nupepa Kuokoa.* Translated by Mary Kawena Pukui. Photocopy of typescript in Library of Hawai'i Institute of Marine Biology, University of Hawai'i, Honolulu, n.d.

Kamakau, Samuel Mānaiakalani. 1976. *The Works of the People of Old: Na Hana a ka Po'e Kahiko.* Translated by Mary Kawena Pukui. Edited by Dorothy B. Barrère. Honolulu, HI: Bishop Museum Press.

———. 1991. *Tales and Traditions of the People of Old: Nā Mo'olelo o ka Po'e Kahiko.* Translated by Mary Kawena Pukui. Edited by Dorothy B. Barrère. Honolulu, HI: Bishop Museum Press.

———. 1992. *Ruling Chiefs of Hawaii.* Translated by Mary Kawena Pukui, Lahilahi Webb, and others. Rev. ed., Honolulu, HI: Kamehameha Schools Press.

Kame'eleihiwa, Lilikalā. 1992. *Native Land and Foreign Desires: Pehea Lā E Pono Ai?* Honolulu, HI: Bishop Museum Press.

———. 1996. *A Legendary Tradition of Kamapua'a: The Hawaiian Pig God.* Honolulu, HI: Bishop Museum Press.

Kane, Herb Kawainui. 1976. *Voyage: The Discovery of Hawaii.* Honolulu, HI: Island Heritage Limited.

———. 1997. *Ancient Hawai'i.* Captain Cook, HI: Kawainui Press.

Kawaharada, Dennis. 1999. "Notes on the Discovery and Settlement of Polynesia." Polynesian Voyaging Society. Accessed October 8, 2015. http://pvs.kcc.hawaii.edu.

Kawelo, Angelo Hiʻilani. 2009. Interview by author, January 29, tape recording, Kāneʻohe, Hawaiʻi.

Kent, Harold Winfield. 1986. *Treasury of Hawaiian Words in One Hundred and One Categories*. Honolulu: Masonic Public Library of Hawaiʻi.

Kneubuhl, Hina. 2008. Interview by author, October 2, email.

Kuoha, Keoni. 2012. Interview by author, phone conversation. Honolulu, Hawaiʻi, September 30.

Larson, Dane. 2002. "The Making of a Surf Ghetto." SurfPulse.com, August 13. Accessed August 9, 2005. http://www.surfpulse.com.

———. 2005. "Is Surfing Etiquette Dead?" SurfPulse.com. Accessed August 9, 2005. http://www.surfpulse.com.

Lefebvre, Henri. 2004. *Rhythmanalysis: Space, Time, and Everyday Life*. New York: Continuum.

Lehrer, Jonah. 2007. *Proust Was a Neuroscientist*. New York: Houghton Mifflin Company.

Lewis, David. 1994. *We, the Navigators: The Ancient Art of Landfinding in the Pacific*. 2nd ed. Honolulu: University of Hawaiʻi Press.

Low, Sam. 2008. "The Ceremony." SamLow.com, September 14. Accessed October 8, 2015. http://www.samlow.com.

Lueras, Leonard. 1984. *Surfing: The Ultimate Pleasure*. Honolulu, HI: Emphasis International.

McGregor, Davianna. 2007. *Nā Kuaʻāina: Living Hawaiian Culture*. Honolulu: University of Hawaiʻi Press.

Meyer, Manulani Aluli. 2001. "Our Own Liberation: Reflections on Hawaiian Epistemology." *Contemporary Pacific* 13 (1): 124–48.

———. 2003. *Hoʻoulu: Our Time of Becoming—Hawaiian Epistemology and Early Writings*. Honolulu, HI: ʻAi Pohaku Press.

Miller, Sunny. 2005. "Summer Lovin." *Water Magazine* 4 (2): 15.

Muneno, Kathy. 2006. "Hawaiʻi Students Navigate Hokuleʻa." *Makai: Ocean Lifestyle Magazine* August, 24–25.

Murphy, Garth. 2004. "Surfing and the Pleasure Principle." *Surfer's Journal* 13 (5): 74–81.

Mykkänen, Juri. 2003. *Inventing Politics: A New Political Anthropology of the Hawaiian Kingdom*. Honolulu: University of Hawaiʻi Press.

Nakuina, Moses Kuaea. 2005. *The Wind Gourd of Laʻamaomao*. Translated by Esther T. Mookini and Sarah Nākoa. Honolulu: University of Hawaiʻi Press.

Nāmakaokeahi, Benjamin K. 2004. *The History of Kanalu Moʻokūʻauhau ʻElua*. Translated and edited by Malcolm Nāea Chun. Honolulu, HI: First People's Productions.

New World Encyclopedia. "Epistemology." Accessed April 4, 2016. http://www.newworldencyclopedia.org/entry/Epistemology.

Ormrod, Joan. 2005. "Endless Summer (1964): Consuming Waves and Surfing the Frontier." *Film and History* 35 (1): 39–51.

Parkinson, Joel. 2008. "Keeping Secretes: Discovery in the Western Pacific." *Water Magazine*, no. 26 (summer): 56–72.

Polynesian Voyaging Society. 2005a. *"History of the Polynesian Voyaging Society: 1973–1998."* S.v. "About PVS Navigation." Polynesian Voyaging Society. Accessed June 15, 2005. http://www.pvs-hawaii.com.

———. 2005b. "Ku Holo La Komohana/Sail On to the Western Sun." Polynesian Voyaging Society. Accessed June 13, 2005. http://www.pvs-hawaii.com.

Provenzano, Renata. 2007. *Kai: Ocean Wisdom from Hawai'i*. Honolulu: Watermark.

Pukui, Mary Kawena. 1949. "Songs (Mele) of Old Ka'u, Hawai'i." *Journal of American Folklore* 62 (265): 247–58.

———. 1983. *'Ōlelo No'eau: Hawaiian Proverbs and Poetical Sayings*. Honolulu, HI: Bishop Museum Press.

Pukui, Mary Kawena, and Alfons L. Korn. 1973. *The Echo of Our Song: Chants and Poems of the Hawaiians*. Honolulu: University of Hawai'i Press.

Pukui, Mary Kawena, and Samuel H. Elbert. 1986. *Hawaiian Dictionary: Hawaiian-English, English-Hawaiian*. Honolulu: University of Hawai'i Press.

Pukui, Mary Kawena, Samuel H. Elbert, and Esther T. Mookini. 1974. *Place Names of Hawaii*. Honolulu: University of Hawai'i Press.

Rancière, Jacques. 1999. *Disagreement: Politics and Philosophy*. Translated by Julie Rose. Minneapolis: University of Minnesota Press.

———. 2004. *The Politics of Aesthetics*. Translated by Gabriel Rockhill. New York: Continuum.

Richardson, Brian W. 2005. *Longitude and Empire: How Captain Cook's Voyages Changed the World*. Toronto: UBC Press.

Rubellite, Kawena Johnson. 1981. *Kumulipo: The Hawaiian Hymn of Creation*. Honolulu, HI: Topgallant.

Said, Edward W. 1993. *Culture and Imperialism*. New York: Vintage Books.

Schultz, Kanekoa Kukea. 2009. Interview by author, January 29, tape recording, Kāne'ohe, Hawai'i.

Shapiro, Michael J. 2000. "National Times and Other Times: Re-Thinking Citizenship." *Cultural Studies* 14 (1): 79–98.

———. 2002. "Bourdieu, the State, and Method." *Review of International Political Economy* 9 (4): 610–18.

———. 2004. *Sovereign Lives: Power in Global Politics*. New York: Routledge.

Silva, Noenoe, 2004. *Aloha Betrayed: Native Hawaiian Resistance to American Colonialism*. Durham, NC: Duke University Press.

———. 2007. "Pele, Hi'iaka, and Haumea: Women and Power in Two Hawaiian Mo'olelo." *Pacific Studies: A Multidisciplinary Journal* 30 (1/2): 159–81.

Smith, Linda Tuhiwai. 1999. *Decolonizing Methodologies: Research and Indigenous Peoples*. New York: Zed Books.

Soguk, Nevzat. 2003. "Incarcerating Travels: Travel Stories, Tourist Orders, and the Politics of the 'Hawai'ian' Paradise." *Tourism and Cultural Change* 1 (1): 29–53.

Sterling, Elspeth P., and Catherine C. Summers, comps. 1978. *Sites of Oahu*. Honolulu, HI: Bishop Museum Press.

Subramani. 2001. "The Oceanic Imaginary." *Contemporary Pacific Journal* 13 (1): 149–62.

Sullivan, Robert. 1999. *Star Waka*. Auckland: Auckland University Press.

Takamine, Vicky-Holt. 2007. Interview by author, January 29, tape recording, Honolulu, Hawai'i.

Teaero, Teweiariki. 2004. *Waa In Storms*. Suva, Fiji: Institute of Pacific Studies, University of the South Pacific.

Teaiwa, Teresia Kieuea. 1994. "bikinis and other s/pacific n/oceans." *Contemporary Pacific* 6 (1): 87–109.

———. 2001. "Militarism, Tourism and the Native: Articulations in Oceania." PhD diss., University of California, Santa Cruz.

———. 2004. Interview by author, November 24, email.

———. 2005. "Native Thoughts: A Pacific Studies Take on Cultural Studies and Diaspora." In *Indigenous Diasporas and Dislocations*, edited by Graham Harvey and Charles D. Thompson, Jr., 15–35. Aldershot, UK: Ashgate Press.

Thaman, Konai Helu. 2003. "Decolonizing Pacific Studies: Indigenous Perspectives, Knowledge, and Wisdom in Higher Education." *Contemporary Pacific* 15 (1): 1–17.

The Oxford English Dictionary, Second Edition. 1989. Prepared by J.A. Simpson and E.S.C. Weiner. Oxford: Clarendon Press.

Thomas, Nicholas. 1994. *Colonialism's Culture: Anthropology, Travel, and Government*. Princeton, NJ: Princeton University Press.

Thompson, Nainoa. 1996. "Master Navigator, Master Teacher." Polynesian Voyaging Society. Accessed October 22, 2007. http://pvs.kcc.hawaii.edu.

———. 2007. "E Ho'i Mau: Honoring the Past, Caring for the Present, Journeying to the Future." *Hūlili: Multidisciplinary Research on Hawaiian Well-Being* 4 (1): 9–34.

———. 2008. Interview by author, December 12, tape recording, Honolulu, Hawai'i.

Timmons, Grady. 1989. *Waikiki Beachboy*. Honolulu, HI: Editions Limited.

Trask, Haunani-Kay. 1994. *Light in the Crevice Never Seen*. Corvallis, OR: Calyx Books.

Tuan, Yi-Fu. 1977. *Space and Place: The Perspective of Experience*. Minneapolis: University of Minnesota Press.

Van, James. 1991. *Ancient Sites of Oahu: A Guide to Hawaiian Archeological Places of Interest*. Honolulu, HI: Bishop Museum Press.

Villela-Petit, Maria. 2010. "Narrative Identity and Ipseity by Paul Ricoeur: From Ricoeur's 'Time and Narrative' to 'Oneself as an Other.'" *Ontology: Online Originals*. Accessed October 8, 2015. http://www.onlineoriginals.com/showitem.asp?itemID=287&articleID=11.

Walker, Isaiah Helekunihi. 2008. "Hui Nalu, Beachboys, and the Surfing Boarder-Lands of Hawai'i." *Contemporary Pacific* (20) 1: 89–113.

Welland, Michael. 2009. *Sand: The Never-Ending Story*. New York: Oxford University Press.

Westervelt, William D. 1910. *Legends of Ma-Ui A Demi-God of Polynesia and of His Mother Hina*. Honolulu: The Hawaiian Gazette Co., Ltd.

———. 1915. *Legends of Old Honolulu*. Boston: George H. Ellis.

Whitman, Walt. 1992. *Leaves of Grass: The Deathbed Edition*. New York: Book-of-the Month Club.

Willis, Koko, and Pali Jae Lee. 1990. *Tales from the Night Rainbow: Moʻolelo o na Pō Mākole*. Honolulu, HI: Night Rainbow.

Wilson, Rob. 1999. "Introduction: Toward Imagining a New Pacific." In *Inside Out: Literature, Culture Politics, and Identity in the New Pacific*, edited by Vilsoni Hereniko and Rob Wilson, 1–16. Lanham, MD: Rowman and Littlefield.

———. 2000. *Reimagining the American Pacific: From South Pacific to Bamboo Ridge and Beyond*. Durham, NC: Duke University Press.

Winner, David. 2000. *Brilliant Orange: The Neurotic Genius of Dutch Soccer*. New York: Overlook Press.

Wood, Houston. 2006. "Three Competing Research Perspectives for Oceania." *Contemporary Pacific* 18 (1): 33–55.

colonization: erasure of place and, 65–72; he'e nalu and effects of, 43–48; indigenous resistance to, 73–78; marginalization of indigenous knowledge and identity and, 3, 8–11, 163–64; ontological time and, 114–19; seascape epistemology and, 20–24; surf tourism and, 5–8, 12–15, 37; travel ideology and, 144–49

conservation. *See* environmental activism

Cook, James, 145–46

Country Feeling Surfboards, 76

cultural sovereignty, colonial deconstruction of, 48–49, 186n.3

Dalai Lama, 165

Daly, Martin, 67–71

Dasein, seascape epistemology and, 28, 122–23

decolonization. *See* postcolonialism

DeLanda, Manuel, 28, 109–11

Deleuze, Gilles, 28, 119, 121–22

DeSoto, Duane, 76, 172–74

deterritorialization, space concepts of Kanaka and, 119–25

Diaz, Vincente, 27, 32, 136–43

Dougherty, Carol, 147

Duke Paoa Kahinu Mokoe Hulikohola Kahanamoku, 13–15, 42

Durham, Jimmie, 56

Earle, Sylvia, 20

economic conditions in Hawai'i: surf tourism and, 70–72; time concepts of Kanaka and, 118–19

Eddie Would Go (Coleman), 62

education in seascape epistemology: *hālau o ke kai* program for, 155–60; *lawai'a* (fishing) programs and, 174–80; Nā Kama Kai project and, 156, 171–74; practice-based/place-based systems for, 161–64; theoretical framework for, 180–82; *wa'a kaulua* voyaging as tool for, 164–71

Elbert, Samuel H., 44

emotion: oceanic literacy and, 82–92; ontological affinity and, 111–13

"Endless Summer (1964): Consuming Waves and Surfing the Frontier" (Ormrod), 54–56

Endless Summer Collection, The (film trilogy), 54–60, 62, 186n.6

English, seascape epistemology in, 30–37

Enlightenment discourse: geography and, 147–49; knowlege and literacy in, 30–31; seascape epistemology and, 96–97

environmental activism: ethics and, 96–101; indigenous surfers involvement in, 75–78; *lawai'a* fishing practices and, 176–80; oceanic literacy and, 95–96

essentialism, indigenous studies and, 185n.1

etak, Carolinean concept of, 137–43

ethics of oceanic literacy, 97–101, 152–54

exploitation films, surfing images in, 53–60

Fernandez, Ramona, 30–31, 92

film images of Hawai'i, surf tourism and, 51–60

Finney, Ben, 48, 98–99

fishing. See *lawai'a* (fishing)

Fluker, Martin, 11–12

Foucault, Michel, 54

Fukushima Daiichi Nuclear Power Plant disaster, 21

genealogy: ancestral fishing knowledge and, 175–80; in Hawaiian culture, 7–11; hierachical order in ancient Hawai'i and, 106–8; oceanic literacy and, 90–92; politics of, 117–19; surfing as connection to, 79–80; voyaging and, 149–52

geography: he'e nalu and, 41–48; human geography and, 82–83; Polynesian geographic rights and, 107; travel ideology and, 146–49

geology, seascape epistemology and, 8–11

Gidget (film), 53

global popularity of surfing, 53–60

global warming, politics of oceanic literacy and, 95–96

grassroots organizations, of indigenous surfers, 74–78

Great Māhele, 11

Groundswell Society, 75

Guattari, Felix, 28, 119, 121

www.ingramcontent.com/pod-product-compliance
Lightning Source LLC
Chambersburg PA
CBHW071102280326
41928CB00051B/2766